STATISTICAL AND INDUCTIVE PROBABILITIES

HUGUES LEBLANC

DOVER PUBLICATIONS, INC.
Mineola, New York

Bibliographical Note

This Dover edition, first published in 2006, is an unabridged republication of the work originally published in 1962 by Prentice-Hall, Inc., Englewood Cliffs, New Jersey.

International Standard Book Number: 0-486-44980-7

Manufactured in the United States of America
Dover Publications, Inc., 31 East 2nd Street, Mineola, N.Y. 11501

TO VIRGINIA

PREFACE

A bitter controversy has been raging among probability theorists ever since the publication of von Mises' "Grundlagen der Wahrscheinlichkeitsrechnung" (1919) and Keynes' *A Treatise on Probability* (1921). Von Mises, as the reader may know, insisted that probabilities measure the relative frequency with which the members of a so-called reference set belong to another set. Keynes, on the other hand, insisted that they measure the extent to which a so-called evidence proposition or, as I shall put it, evidence sentence supports another sentence.

An attempt is made here to settle a dispute which cannot but be prejudicial to philosophy of science and, by rebound, to science itself. In the opening chapter I review, for the reader's convenience, the essentials of sentence theory and of set theory. In the next chapter I study in some detail those probabilities, often called *statistical probabilities*, which are currently allotted to sets by von Mises' followers. In the third chapter I show that statistical probabilities may be passed on to sentences and thereupon become truth-values of one sort or another. In the last chapter I study those probabilities, often called *inductive probabilities*, which Keynes' followers currently allot to sentences, and show that they may be reinterpreted as estimates of truth-values. It should thus appear, by the close of the book, that both statistical and inductive probabilities may be treated as sentence-theoretic measurements and that the latter qualify as estimates of the former.

Anxious as I may be to bring statistical and inductive probabilities together, I remain alert, I hope, to their distinctive characteristics. Statistical probabilities, as I see it, betoken some indefiniteness in the sentences to which they are allotted—indefiniteness as to the exact subject matter of those sentences. Inductive probabilities, on the other hand, betoken some uncertainty on our part—uncertainty as to the exact truth-value of the sentences to which they are allotted. This contrast between indefiniteness as to subject matter and uncertainty as to truth-value is stressed and exploited at critical points in the text.

An attempt is also made to sketch the variety of ways in which the probability of a pair of sets or of a pair of sentences may be reckoned. In much of the literature, to be sure, the members of a so-called probability set are all weighted alike. There are, nonetheless, infinitely many ways of weighting those members, and the probability of a pair of sets may vary, as will appear in Chapter 2, from one allotment of weights to another. Similarly, there are infinitely many ways of weighting the so-called individual constants or the so-called state-descriptions of a language, and the probability (statistical in one case, inductive in the other) of a pair of sentences may vary, as will appear in Chapters 3 and 4, from one allotment of weights to another.

Finally, an attempt is made to study the inferential uses to which statistical and inductive probabilities may respectively be put. I show, for example, that the so-called coefficient of statistical reliability of an inductive inference can be reckoned by means of statistical probabilities, a point sometimes overlooked by writers of Keynes' persuasion. I also show that the coefficient in question can be estimated by means of inductive probabilities when the statistical data needed for reckoning it are wanting. I finally note that the hypergeometric theorem, upon which so many inductive inferences are founded, holds for all of the inductive probability functions considered in Chapter 4. The material is offered as a preliminary contribution to inductive logic.

As a rule, studies in probability theory make for arduous reading, and mine is no exception. I have, however, tried to smooth the reader's path. A summary of things to come heads each chapter; illustrations accompany most definitions and theorems; footnotes, collected for the reader's convenience at the end of each chapter, elucidate various technicalities and bring him up-to-date on the literature of the subject; finally, a roster of the logical and mathematical symbols employed in the text is supplied on page 140.

While I was writing this book, I had the good fortune of securing the assistance of numerous colleagues who read early drafts and offered invaluable criticisms and suggestions. My thanks must go, in this regard, to Rudolf Carnap (The University of California at Los Angeles), Theodore Hailperin (Lehigh University), Hans Hermes (The University of Münster), Stig Kanger (The University of Stockholm), Robert McNaughton (The University of Pennsylvania), Ernest Nagel (Columbia University), Walter Oberschelp (The University of Münster), Joseph Ullian (The University of Chicago), William A. Wisdom (The Pennsylvania State University), and Georg H. von Wright (The University of Helsingfors). I also thank Robert B. Burlin (Bryn Mawr College), Donald J. Hillman (Lehigh University), and Miss Beatrice Yamasaki (Los Angeles State College) for reading the galleys with me; Messrs James Gord, Christoph E. Schweitzer, and Clifford G. Walters for attending to various clerical matters; and Richard W. Hansen, George Chien, and Herbert Nolan of Prentice-Hall, Inc. for placing such faith in the book and showing its author so much patience.

I am indebted to the editors of numerous journals here and abroad, *The British Journal for the Philosophy of Science*, *The Journal of Philosophy*, *The Journal of Symbolic Logic*, *Notre Dame Journal of Formal Logic*, *Philosophical Studies*, *Philosophy of Science*, *Philosophy and Phenomenological Research*, and *Revue Philosophique de Louvain*, who have given me repeated opportunities to air some of my views on probability; and to various institutions and learned groups I have been privileged to address on the subject, among them Bryn Mawr College, Le Centre National Belge de recherches de Logique, Georgetown University (Institute in the Philosophical Foundations of Mathematics and Physics, 1961), The Pennsylvania State University, Stanford University (1960 International Congress for Logic, Methodology and Philosophy of Science), The University of California at Los Angeles, Wesleyan University (Wesleyan Conference on Induction, 1961), The American Philosophical Association, The Association for Symbolic Logic, and The Fullerton Club.

I am further indebted to Bryn Mawr College for awarding me its Eugenia Chase Guild Fellowship and granting me a sabbatical leave for the academic year 1958–1959, to The American Philosophical Society for awarding me research grants in 1956 and 1958, and to Bryn Mawr College again for defraying through the offices of its Committee on the Madge D. Miller Fund some of the clerical expense occasioned by the book.

Bryn Mawr College Hugues Leblanc

CONTENTS

1 THE LANGUAGES L 1

1. The vocabulary of the languages L 2
2. The grammar of the languages L 4
3. Designation, extension, and truth in the languages L 9
4. Logical truth, implication, and equivalence in the languages L 15
5. Set theory and the language L 21

Notes 29

2 STATISTICAL PROBABILITIES: PART ONE 33

6. Statistical probabilities allotted to sets 34
7. Comments and illustrations 38
8. Absolute probabilities, probability sets, and weights 47
9. Random functions 51
10. Random sampling and attendant distributions 53
11. Statistical probabilities estimated by means of sample soundings 57

Notes 61

3 STATISTICAL PROBABILITIES: PART TWO 67

12. Sets, sentences, and statistical probabilities 68
13. Statistical probabilities allotted to the sentences of the languages *L* 71
14. Statistical probabilities as truth-values 76
15. The relative frequency allotment 78
16. The statistical probability of closed sentences 85
17. A first look at inductive inferences 89

Notes 93

4 INDUCTIVE PROBABILITIES 97

18. Inductive probabilities allotted to the sentences of the languages *L* 98
19. Variants of the standard allotments (I) 104
20. Variants of the standard allotments (II) 107
21. Inductive probabilities as estimates of truth-values 112
22. Logical falsehoods and logical truths qua evidence sentences 122
23. Personal versus inductive probabilities 126
24. A second look at inductive inferences 129

Notes 134

LIST OF SYMBOLS 140

BIBLIOGRAPHICAL REFERENCES 141

INDEX OF AUTHORS 145

INDEX OF MATTERS 146

1 THE LANGUAGES L

In this first chapter I construct a family of languages, the languages L, which is to consist of a master language, called L^{∞}, and various fragments or sublanguages of L^{∞}, respectively called L^1, L^2, L^3, and so on ad infinitum. The sets allotted probabilities in Chapter 2, for one thing, can all be mentioned in the languages L; the sentences allotted probabilities in Chapters 3 and 4, for another, will all be sentences of these languages. First, I study the vocabulary and the grammar of the languages L (Sections 1–2); then, I study the interpretation to be placed upon the languages L (Sections 3–4); finally, I graft onto the languages L a set theory of sorts (Section 5). The material, made up in equal parts of definitions and marginal comments thereon, should be familiar to most; it may, however, be novel to some, who will perhaps welcome a briefing on the logic of sentences and the logic of sets.[1]

1. THE VOCABULARY OF THE LANGUAGES L

My first group of definitions deals with the vocabulary of the languages L. The so-called primitive signs of L^∞ are listed in D1.1; the primitive signs of the sublanguages of L^∞ are listed in D1.2; the order in which some of these signs are listed or presumed to be listed in D1.1–2 is, for technical convenience, given a name in D1.3. Further signs, to be known as the defined signs of L^∞ and its sublanguages, will be supplied in Sections 2 and 5.

D1.1. *The primitive signs of L^∞ consist of the following:*

(a) *The two connectives '$\sim$' and '$\supset$';*

(b) *The quantifier letter '$\forall$';*

(c) *The identity sign '$=$';*

(d) *The comma ',';*

(e) *The two parentheses '(' and ')';*

(f) *A finite set of predicates (each identified as a one-place predicate or a two-place predicate or a three-place predicate, and so on);*

(g) *A denumerably infinite set of individual constants;*[2]

(h) *A denumerably infinite set of individual variables:*

$$w, x, y, z, w', x', y', z', \cdots .$$

D1.2. *For each N from 1 on, the primitive signs of the sublanguage L^N of L^∞ consist of the following:*

(a)–(f) *The various signs listed in D1.1(a)–(f);*

(g) *The first N individual constants listed in D1.1(g);*

(h) *The individual variables listed in D1.1(h).*

D1.3. (a) *The order in which the individual constants of L^∞ and its sublanguages are presumed to be listed in D1.1(g) and D1.2(g) is the alphabetical order of those constants.*

(b) *The order in which the individual variables of L^∞ and its sublanguages are listed in D1.1(h) and D1.2(h) is the alphabetical order of those variables.*[3]

A word of explanation on each one of clauses D1.1(a)–(h) may be in order. The connective '$\sim$' in D1.1(a) may be read 'It is not the case that' or, when the occasion warrants, 'Not.' The connective

'$\supset$' in D1.1(a) may be read 'If $\cdots$, then.' The quantifier letter '$\forall$' in D1.1(b) may be read 'For any.' The identity sign '$=$' in D1.1(c) may be read 'is identical with,' 'is the same as,' or, when the occasion warrants, 'is.' The comma ',' and the two parentheses '(' and ')' in D1.1(d)–(e) are punctuation signs of a sort; the role played by ',' should be clear from D2.2(a) below, the one played by '(' and ')' clear from D2.2(c)–(e) and the explanations appended thereto. The predicates presumed to be listed in D1.1(f) are expressions like 'won the 1948 presidential election,' 'is one of . . .'s satellites,' or 'extends from . . . to,' which, once supplied at appropriate places with nouns like 'Truman,' 'Enceladus,' 'Saturn,' 'Canada,' 'the Atlantic Ocean,' or 'the Pacific Ocean,' yield what are called closed sentences:[4] 'Truman won the 1948 presidential election,' 'Enceladus is one of Saturn's satellites,' 'Canada extends from the Atlantic Ocean to the Pacific Ocean.' Among the expressions in question, those which must be supplied with one noun (and one only) to yield a closed sentence are called one-place predicates; those which must be supplied with two nouns (and two only) to yield such a sentence are called two-place predicates; and so on. The individual constants presumed to be listed in D1.1(g) are nouns like the above 'Truman,' 'Enceladus,' and 'Saturn'; they are meant to go with the predicates and the identity sign '$=$' of the languages L; they are also meant to designate among themselves all the individuals making up what are called the universes of discourse of the languages L. Finally, the individual variables listed in D1.1(h) are to fill various roles: they will go (as individual constants do) with the predicates and the identity sign '$=$' of the languages L and yield what are called open sentences such as 'Peter knows w,' '(w is a man) $\supset$ (w is mortal),' '$w = w$,' and so on; they will serve with the aid of the quantifier letter '$\forall$' to turn such open sentences into closed ones like '$(\forall w)$(Peter knows w),' '$(\forall w)$((w is a man) $\supset$ (w is mortal)),' and $(\forall w)(w = w)$';[5] and they will indiscriminately refer to or, as the technical phrase goes, range over the individuals designated in the languages L by the individual constants of those languages.

As the reader will have noticed, I do not produce the finitely many predicates of L^∞; I simply assume that L^∞ is fitted with such predicates. Neither do I produce the infinitely many individual constants of L^∞; I simply assume that L^∞ is fitted with such constants.[6] The procedure, standard in studies of this kind, makes for greater generality. It also gives one greater leeway when it comes to illustrating sundry points about L^∞ and its sublanguages.

So much for D1.1. According to D1.2 the sublanguages L^1, L^2, L^3, and so on, of L^∞ are like L^∞ except for boasting only the first one, the first two, the first three, and so on, of the infinitely many individual constants of L^∞. The sublanguages in question will serve various technical purposes; they will also be of theoretical interest in themselves. The universe of discourse of L^1 will, for example, be of size 1, that of L^2 of size 2, that of L^3 of size 3, and so on, as opposed to the universe of discouse of L^∞, which will be denumerably infinite in size or, as the matter is often put, of size $\aleph_0$.[7]

2. THE GRAMMAR OF THE LANGUAGES L

My next group of definitions deals with the grammar of (the languages) *L*.[8] First, I single out from among all the sequences of primitive signs of *L* those which are to be known as the expressions of *L* (D2.1). Then, I single out from among all the expressions of *L* those which are to be known as the sentences of *L* (D2.2). Next, I sort the sentences of *L* into two groups: the closed sentences of *L* (D2.4) and the open sentences of *L* (D2.5).[9] Finally, I define what I understand by an instance in *L* of a sentence of *L* (D2.6). The sorting of the sentences of *L* into closed sentences and open ones calls for an auxiliary notion, that of a free individual variable of *L*. I define the notion in D2.3.

The capitals with which D2.2, D2.6, and so on, fairly bristle are so-called metalinguistic variables. Four of those variables, '*P*,' '*Q*,' '*R*,' and '*S*,' range over the sentences of *L*; four more, '*W*,' '*X*,' '*Y*,' and '*Z*,' range over the individual signs (that is, the individual constants and the individual variables) of *L*; and a ninth one, '*G*,' ranges over the predicates of *L*. Finally, sequences of metalinguistic variables and signs of *L* range over the results of substituting for the variables in the sequences the various expressions of *L* over which the variables range. In D2.2(a), for example, '$G(W_1, W_2, \cdots, W_n)$' ranges over the results of substituting a predicate of L for 'G' and individual signs of L for 'W_1,' 'W_2,' $\cdots$, and 'W_n' in '$G(W_1, W_2, \cdots, W_n)$'; in D2.2(d), '$(P) \supset (Q)$' ranges over the results of substituting sentences of L for 'P' and 'Q' in '$(P) \supset (Q)$'; and so on.

D2.1. *An expression of L is a finite sequence of primitive signs of L.*

D2.2. (a) *$G(W_1, W_2, \cdots, W_n)$, where G is an n-place ($n \geq 1$) predicate of L and $W_1, W_2, \cdots$, and W_n are n individual signs of L, is a sentence of L;*

(b) *$W = X$, where W and X are two individual signs of L, is a sentence of L;*

(c) *If P is a sentence of L, then so is $\sim (P)$;*

(d) *If P and Q are two sentences of L, then so is $(P) \supset (Q)$;*

(e) *If P is a sentence of L, then so is $(\forall W)(P)$, where W is an individual variable of L;*

(f) *No expression of L is a sentence of L unless its being so follows from (a)–(e).*[10]

D2.3. (a) *An occurrence of an individual variable W of L in a sentence P of L is bound in P if it is in a subsequence $(\forall W)(Q)$ of P;*[11]

(b) *An occurrence of an individual variable W of L in a sentence P of L is free in P if it is not bound in P;*

(c) *Any occurrence of an individual constant W of L in a sentence P of L is free in P;*

(d) *An individual sign W of L is bound in a sentence P of L if at least one occurrence of W in P is bound in P;*

(e) *An individual sign W of L is free in a sentence P of L if at least one occurrence of W in P is free in P.*

D2.4. *A closed sentence of L is a sentence of L in which no individual variable of L is free.*

D2.5. *An open sentence of L is a sentence of L which is not closed.*

D2.6. (a) *Let P be a closed sentence of L. Then P is the instance of P in L.*

(b) *Let P be an open sentence of L; let $W_1, W_2, \cdots$, and W_n be the n ($n \geq 1$) individual variables of L which are free in P; and let P^* be* *like P except for containing, for each i from 1 to n, occurrences of an individual constant of L at all the places where P contains free occurrences of W_i. Then P* is an instance of P in L.*

Among the expressions pronounced sentences in D2.2(a) are to be found, for example, 'Cervantes is a Spanish writer,' 'Mary just came back from w,' or 'w lies between x and y,' which I throw for the occasion into the form 'Is a Spanish writer(Cervantes),' 'Just came back from(Mary, w),' and 'Lies between and(w,x,y),' and so on. Among those pronounced sentences in D2.2(b) are to be found, for example, 'The Morning Star = the Morning Star,' 'w = the 35th President of the U.S.A.,' or '$w = x$,' expressions commonly called identities. Among those pronounced sentences in D2.2(c) are to be found, for example, '∼ (Philadelphia = the capital of Pennsylvania),' '∼ ((Has read(w,x)) ⊃ (Remembers(w,x))),' or '∼ (($\forall w$)(Floats on water(w))),' expressions commonly called negations. Among those pronounced sentences in D2.2(d) are to be found, for example, '(Votes Democratic(Harry)) ⊃ (Votes Republican(John)),' '((($\forall w$)($w = w$)) ⊃ ($x = x$),' or '(Likes(Ann,w)) ⊃ (($\forall x$)((Is a friend of(x,w)) ⊃ (Likes(Ann,x)))),' expressions commonly called conditionals. Among those pronounced sentences in D2.2(e) are to be found, for example, '($\forall w$)($w = w$),' '($\forall w$)((Is a swan(w)) ⊃ (Is white(w))),' or '($\forall w$)(($\forall x$)((Is the father of(w,x)) ⊃ (∼ (Is the father of(x,w))))),' expressions commonly called universal sentences.

The parentheses that officially go around P in a negation $\sim(P)$, around P and Q in a conditional $(P) \supset (Q)$, and around P in a universal sentence $(\forall W)(P)$ are indispensable when P in the first case, P or Q in the second, and P in the third are conditionals. Without them, indeed, we could no longer tell $\sim(P \supset Q)$ from $\sim P \supset Q$, $(P \supset Q) \supset R$ from $P \supset (Q \supset R)$, nor $(\forall W)(P \supset Q)$ from $(\forall W)P \supset Q$. Otherwise, the parentheses in question may be dispensed with, and I shall frequently do so.

A few examples should shed light on the distinctions drawn in D2.3. Consider the three sentences:

$$(\forall w)(\text{Is colored}(w)) \supset \text{Is extended}(w), \quad (1)$$

$$(\forall w)(\text{Is colored}(w) \supset \text{Is extended}(w)), \qquad (2)$$

and

$$\text{Is colored}(w) \supset \text{Is extended}(w). \qquad (3)$$

The first two occurrences of 'w' in (1) are in a subsequence of (1) which opens with '$(\forall w)$'—the subsequence '$(\forall w)$(Is colored(w))'—and hence are bound in (1); the third occurrence of 'w' in (1), on the other hand, is free in (1). All three occurrences of 'w' in (2) are in a subsequence of (2) which opens with '$(\forall w)$'—the subsequence (2)—and hence are bound in (2). Finally, both occurrences of 'w' in (3) are free in (3). As for 'w' itself, it is both bound and free in (1), bound (and bound only) in (2), and free (and free only) in (3). Occurrences of individual constants, and individual constants themselves, are pronounced free in D2.3 for the sake of convenience.

According to D2.4, 'Is west of(Toronto,Montreal) $\supset \sim$ Is west of(Montreal,Toronto),' '$(\forall x)$(Is west of(Toronto,x) $\supset \sim$ Is west of (x,Toronto)),' and '$(\forall w)(\forall x)$(Is west of(w,x) $\supset \sim$ Is west of(x,w))' are closed sentences, no individual variable being free in any one of the three sentences. According to D2.4–5, on the other hand, 'Is west of(w,x) $\supset \sim$ Is west of(x,w),' 'Is west of(w,Montreal) $\supset \sim$ Is west of(Montreal,w),' and '$(\forall w)$(Is west of(w,x) $\supset \sim$ Is west of(x,w))' are open sentences, 'w' and 'x' being free in the first sentence, 'w' in the second, and 'x' in the third.

There are clearly two ways of turning an open sentence into a closed one. The first is to put the sentence in parentheses and preface the resulting expression with one universal quantifier, that is, one expression of the form $(\forall W)$ per individual variable W which is free in the sentence. We can, for example, turn the open sentence 'Weighs as much as(w,x) $\supset$ Weighs as much as(x,w)' into a closed one by putting it in parentheses and prefacing the resulting expression with '$(\forall w)(\forall x)$'; the outcome will read: '$(\forall w)(\forall x)$(Weighs as much as(w,x) $\supset$ Weighs as much as(x,w)).' The other way is to substitute (occurrences of) individual constants for the free occurrences of the various individual variables which are free in the sentence. We can turn the open sentence 'Excels at(Jack,w) $\supset \sim (\forall w) \sim$ Excels at(Jack,w),' for example, into a closed one by substituting 'poker' (or any other individual constant we please) for the initial occurrence of 'w' in the sentence; the outcome will read: 'Excels at(Jack, poker) $\supset \sim (\forall w) \sim$ Excels at(Jack,w).'

The outcomes of thus substituting individual constants of L for the free occurrences of the individual variables of L that are free in an open sentence of L are what I call the instances of the sentence in L (D2.6). An open sentence of L in which n individual variables of L, say W_1, W_2, $\cdots$, and W_n, are free has in a sublanguage L^N of L^∞ (or, as I shall often put it, in L^N) N^n instances. Note for proof that (1) W_1 can be substituted for in N different ways; (2) for each one of the N ways in which W_1 can be substituted for, W_2 can be substituted for in N different ways; (3) for each one of the N^2 ways in which W_1 and W_2 can be substituted for, W_3 can be substituted for in N different ways; and so on. In L^∞ the same sentence has $\aleph_0{}^n$ instances and hence, $\aleph_0{}^n$ being equal to $\aleph_0$, $\aleph_0$ instances.[12] It proves convenient to extend the classical notion of an instance and allow, as I did in D2.6(a), a closed sentence of L to serve as its own instance in L.

Before closing this section, I graft onto L five so-called defined signs. The first three are sentence connectives, namely, '&,' 'v,' and '≡,' to be respectively read 'and,' 'or' (more explicitly, 'and/or'), and 'if and only if.' The fourth one is a quantifier letter, namely, '∃,' to be read 'For some' or 'There exists at least one $\cdots$ such that.' The fifth one is a predicate, namely, 'D,' which, once supplied with two or more individual constants or variables, yields sentences like 'D(w,x),' 'D(w,x,y),' and so on. These sentences may be read 'w and x are distinct from each other,' 'w, x, and y are distinct from one another,' and so on. All five signs are grafted onto L by means of what we call contextual definitions. D2.7, for example, authorizes us to treat $P \,\&\, Q$ as a rewrite in L of the sentence $\sim (P \supset \sim Q)$; D2.10 to treat $(\exists W)P$ as a rewrite in L of the sentence $\sim (\forall W) \sim P$; and D2.11 to treat $\mathsf{D}(W_1, W_2, W_3)$ as a rewrite in L of the sentence $\sim W_1 = W_2 \,\&\, (\sim W_3 = W_1 \,\&\, \sim W_3 = W_2)$.[13]

D2.7. *(P) & (Q) is defined as* $\sim (P \supset \sim Q)$.

D2.8. *(P) v (Q) is defined as* $\sim P \supset Q$.

D2.9. *(P) ≡ (Q) is defined as* $(P \supset Q) \,\&\, (Q \supset P)$.

D2.10. $(\exists W)(P)$ *is defined as* $\sim (\forall W) \sim P$.

D2.11. **(a)** $\mathsf{D}(W_1, W_2)$ *is defined as* $\sim W_1 = W_2$;

(b) $\mathsf{D}(W_1, W_2, \cdots, W_{n+1})$ *is defined as* $\mathsf{D}(W_1, W_2, \cdots, W_n)$ & $((\cdots(\mathsf{D}(W_{n+1}, W_1)$ & $\mathsf{D}(W_{n+1}, W_2))$ & $\cdots)$ & $\mathsf{D}(W_{n+1}, W_n))$, *where* $n \geq 2$.[14]

3. DESIGNATION, EXTENSION, AND TRUTH IN THE LANGUAGES L

Several matters must be attended to when it comes to interpreting a language: specifying what its individual variables, its individual constants, and its predicates refer to; laying down the conditions under which its closed sentences are true or false; and so on. The first one is usually attended to as follows. Given a language, say L', one or more individuals are first singled out as the range of values in L' of the individual variables of L' or, simply, as the universe of discourse of L'. A member of the universe in question is next assigned to each individual constant of L' as the individual designated in L' by the constant or, more briefly, as the designation in L' of the constant. A set of members of the universe in question is finally assigned to each one-place predicate of L' as the set of the individuals to which the one-place predicate applies in L' or, more briefly, as the extension in L' of the one-place predicate; a set of pairs of members assigned to each two-place predicate of L' as the extension in L' of the two-place predicate; a set of triples of members assigned to each three-place predicate of L' as the extension in L' of the three-place predicate; and so on.

Two things may happen (and are sometimes bound to happen) under the scheme above: (1) a given member of the universe of discourse of L' may come to be designated in L' by two or more individual constants of L'; (2) another member of that universe may not come to be designated in L' by any individual constant of L'. In many contexts it is preferable to leave these two possibilities open; here, however, I choose to exclude them.

Using then another (though related) procedure, I shall presume that a first individual has been assigned to the first individual constant of L^∞ as its designation in L^∞; a second individual, distinct from the first, assigned to the second individual constant of L^∞ as its designation in L^∞; a third individual, distinct from the first two, assigned to the third individual constant of L^∞ as its designation in L^∞; and so on ad infinitum. I shall next presume that a set of those infinitely many individuals has been assigned to each one-place predicate of L^∞ as its extension in L^∞; a set of pairs of them assigned to each two-place predicate of L^∞ as its extension in L^∞; a set of triples of them assigned to each three-place predicate of L^∞ as its extension in L^∞; and so on.

As for the sublanguages of L^∞, each individual constant of L^N ($N = 1, 2, 3, \cdots$) will have as its designation in L^N the individual it designates in L^∞, and each predicate of L^N will have as its extension in L^N the set of those among the individuals designated in L^N by the individual constants of L^N, or those among the pairs of such individuals, or those among the triples of such individuals, and so on, that belong to its extension in L^∞.

It will follow from my account of things here and from definition D3.2 below that the individuals designated in L^∞ by the individual constants of L^∞ make up the entire universe of discourse of L^∞. It will follow by the same token that every member of the universe of discourse of L^∞ is designated in L^∞ by at most one and at least one individual constant of L^∞. The same will hold true of each one of the sublanguages of L^∞.

The word 'individual' which popped up again and again in the last five paragraphs may cover, by the way, anything the reader pleases: things animal, vegetable, or mineral (as in Twenty Questions), sets, relations, numbers, events, and so on. The individual constants of L may thus be nouns like the 'Truman' and 'Enceladus' of Section 1; they may also be nouns like 'the United Nations,' 'parenthood,' 'Ann's semester average,' or 'World War II.' Similarly, the predicates of L may be expressions like the 'won the 1948 presidential election' and 'is one of ··· 's satellites,' of Section 1; they may also be expressions like 'sponsored ··· 's admission to' (as in 'France sponsored Togoland's admission to the United Nations'), 'is a trying responsibility' (as in 'Parenthood is a trying responsibility'), 'falls short of 83' (as in 'Ann's semester average falls short of 83'), or 'was sparked by' (as in 'World War Two was sparked by the invasion of Poland').

D3.1. (a) *Clauses of the following kind:*

(a1) W_1 *has* $---_1$ *as its designation in* L^∞,
(a2) W_2 *has* $---_2$ *as its designation in* L^∞,
.
.
(aN) W_N *has* $---_N$ *as its designation in* L^∞,
.
.

are presumed to be on hand, where $W_1, W_2, \cdots, W_N, \cdots$, *are in alphabetical order the various individual constants of* L^∞ *and the blanks* '$---_1$,' '$---_2$,' $\cdots$, '$---_N$,' $\cdots$, *are filled with names of individuals distinct from one another.*

(b) *Let* $\vee^n_{L^\infty}$ $(n \geq 1)$ *be the set of all the sequences made up of any n of the individuals under (a).*[15] *For each predicate G of* L^∞ *a clause of the kind:*

$$G \text{ has } --- \text{ as its extension in } L^\infty,$$

is presumed to be on hand, where '---' *is filled with a name of a subset of* $\vee^1_{L^\infty}$ *in case G is a one-place predicate, of* $\vee^2_{L^\infty}$ *in case G is a two-place predicate, of* $\vee^3_{L^\infty}$ *in case G is a three-place predicate, and so on.*[16]

(c) *Let* W_i $(i \leq N)$ *be the i-th individual constant of* L^N *and let* $---_i$ *be the individual mentioned in (ai). Then* W_i *has* $---_i$ *as its designation in* L^N.

(d) *Let G be an n-place* $(n \geq 1)$ *predicate of* L^N*; let* $\vee^n_{L^N}$ *be the set of all the sequences made up of any n of the individuals mentioned in (a1)–(aN); let* $---^\infty$ *be the extension of G in* L^∞*; and let* $---^N$ *be the set of all the members of* $---^\infty$ *which belong to* $\vee^n_{L^N}$. *Then G has* $---^N$ *as its extension in* L^N.

Two remarks are in order.

(1) The reader may draft any individual he pleases to serve as the designation in L^∞ of a given individual constant W of L^∞, and any subset of $\vee^n_{L^\infty}$ he pleases (the null set included)[17] to serve as the extension in L^∞ of a given n-place predicate G of L^∞. He would do well, however, if he borrows the individual constant or the predicate in question from every day language, to assign it the designation

or the extension it already has in everyday language; misunderstandings that he would be hard put to dispel might otherwise arise.

(2) The reader may fill the blank in

W has - - - as its designation in L^∞

with any name he pleases of the individual which is to serve as the designation of *W* in L^∞, be it *W* itself or any other name of this individual; he may likewise fill the blank in

G has - - - as its extension in L^∞

with any name he pleases of the subset of $\vee^n_{L^\infty}$ which is to serve as the extension of *G* in L^∞, be it a list of the members of the subset or a mere description of them. A sample clause D3.1(a) might thus read:

'Hemingway' has Hemingway as its designation in L^∞*;*

it might also read:

'Hemingway' has the author of The Sun Also Rises *as its designation in L;*

and so on. A sample clause D3.1(b) might likewise read:

'Was married to Hemingway' has as its extension in L^∞ *the set consisting of the individuals designated in* L^∞ *by the individual constants* W_1, W_{13}, W_{27}, *and* W_{75} *of* L^∞,

where W_1, W_{13}, W_{27}, and W_{75} are presumed for the occasion to designate in L^∞ the various wives of Hemingway; it might also read:

'Was married to Hemingway' has as its extension in L^∞ *the set consisting of the various wives of Hemingway,*

where the various wives of Hemingway are presumed for the occasion to belong to $\vee^1_{L^\infty}$; and so on.[18]

I am now ready to state the conditions under which a closed sentence of *L* is to be true in *L*. Condition D3.2(e) insures, along with D3.1, that the individuals designated in *L* by the individual constants of *L* make up the entire universe of discourse of *L*. Note indeed that if the said individuals made up but part of the said universe, the instances in *L* of an open sentence *P* of *L* could all be true in L without the corresponding universal sentence $(\forall W)P$ of L being also true in L.

D3.2. (a) *If G is an n-place ($n \geq 1$) predicate of L, $W_1, W_2, \cdots$, and W_n are n individual constants of L, and the sequence made up of the n individuals respectively designated in L by $W_1, W_2, \cdots$, and W_n (in that order) belongs to the extension of G in L, then $G(W_1, W_2, \cdots, W_n)$ is true in L;*

(b) *If W and X are two individual constants of L and the individuals respectively designated in L by W and X are the same, then $W = X$ is true in L;*[19]

(c) *If P is a closed sentence of L and P is not true in L, then $\sim P$ is true in L;*

(d) *If P and Q are two closed sentences of L and if P is not true in L and/or Q is true in L, then $P \supset Q$ is true in L;*

(e) *If no individual variable of L distinct from W is free in P and every instance of P in L is true in L, then $(\forall W)P$ is true in L;*[20]

(f) *No closed sentence of L is true in L unless its being so follows from (a)–(e).*[21]

Consider, for illustration's sake, the closed sentence

$$(\forall w)(\text{Is a committee member}(w) \supset (\sim w = \text{Peter} \supset \text{Voted for}(w,\text{Harry}))). \quad (1)$$

(1) will be true if every instance of

$$\text{Is a committee member}(w) \supset (\sim w = \text{Peter} \supset \text{Voted for}(w,\text{Harry})) \quad (2)$$

is true. Among the instances of (2),

$$\text{Is a committee member(Jim)} \supset (\sim \text{Jim} = \text{Peter} \supset \text{Voted for(Jim,Harry)}), \quad (2')$$

for example, will be true if

$$\text{Is a committee member(Jim)} \quad (3)$$

is not true or

$$\sim \text{Jim} = \text{Peter} \supset \text{Voted for(Jim,Harry)} \quad (4)$$

is true. (3) will be true if the individual designated by 'Jim' belongs to the extension of 'Is a committee member.' (4) will be true if

$$\sim \text{Jim} = \text{Peter} \tag{5}$$

is not true or

$$\text{Voted for(Jim,Harry)} \tag{6}$$

is true. (5) will be true if

$$\text{Jim} = \text{Peter} \tag{7}$$

is not true. Finally, (6) will be true if the pair made up of the two individuals respectively designated by 'Jim' and 'Harry' belongs to the extension of 'Voted for,' and (7) will be true if the two individuals respectively designated by 'Jim' and 'Harry' are the same (a possibility precluded, of course, by D3.1).

I next state the one condition under which a closed sentence of L is to be false in L and introduce a function, the so-called truth-value function, whose value **Tv**(P) for a closed sentence P of L is to be 1 if P is true in L, 0 if P is false in L.

D3.3. *A closed sentence of L is false in L if it is not true in L.*

D3.4. *Let P be a closed sentence of L.*
(a) *If P is true in L, then* **Tv**(P) *equals 1;*
(b) *If P is not true in L, then* **Tv**(P) *equals 0.*

I shall occasionally write '$\mathbf{Tv}^{\infty}(P)$' when I wish to emphasize that P is a closed sentence of L^{∞}, '$\mathbf{Tv}^{N}(P)$' when I wish to emphasize that P is a closed sentence of L^{N}. In Chapters 3 and 4 I shall also occasionally refer to sums of truth-values and to limits of sums of truth-values as truth-values in a generalized sense of the word.

An open sentence of L can neither be true nor be false in L. It can, however, be satisfied in L, as the phrase goes, by certain assignments of members of the universe of discourse of L to the individual variables of L which are free in the sentence.

D3.5. *Let* $\vee^1_L$ *be the set of all the individuals designated in L by the individual constants of L; let P be an open sentence of L; let* $W_1, W_2, \cdots,$ *and* W_n *be in alphabetical order the n* ($n \geq 1$) *individual variables of L which are free in P; let* **Asst** *be a given assignment of members of* $\vee^1_L$ *to* $W_1, W_2, \cdots,$ *and* W_n*; let* $X_1, X_2, \cdots,$ *and* X_n *be the n individual constants of L which designate in L the members of* $\vee^1_L$ *respectively assigned by* **Asst** *to* $W_1, W_2, \cdots,$ *and* W_n*; and let P′ be like P except* *for containing, for each i from 1 to n, occurrences of X_i at all the places where P contains free occurrences of W_i. Then P is satisfied in L by* **Asst** *if P′ is true in L.*[22]

For example, let $\vee_L^1$ contain (among other individuals) Lenin and Marx; let P be the open sentence 'Is a disciple of(w,x)'; let $\mathbf{Asst}_1$ assign Lenin to 'w' and Marx to 'x'; and let $\mathbf{Asst}_2$ assign Marx to 'w' and Lenin to 'x.' Since the pair made up of the two individuals respectively designated in L by 'Lenin' and 'Marx' belongs to the extension of 'Is a disciple of' in L when and only when Lenin comes first in the pair and Marx second, P will be satisfied in L by $\mathbf{Asst}_1$, but not by $\mathbf{Asst}_2$.

It should be clear from D3.5 and the remarks appended to D2.6 that members of $\vee_{L^N}^1$ can be assigned to the n individual variables of L^N which are free in an open sentence of L^N in N^n different ways and that members of $\vee_{L^\infty}^1$ can be assigned to the n individual variables of L^∞ which are free in an open sentence of L^∞ in $\aleph_0$ different ways. It should also be clear from D3.5 and D3.2–3 that a closed sentence P of L is true in L if and only if $W = W \supset P$, where W is an individual variable of L, is satisfied in L by every possible assignment of a member of $\vee_L^1$ to W; false in L, on the other hand, if and only if $W = W \supset P$ is satisfied in L by no such assignment. Both facts will prove useful later.

4. LOGICAL TRUTH, IMPLICATION, AND EQUIVALENCE IN THE LANGUAGES L

Compare the three closed sentences:

De Gaulle is French, (1)

De Gaulle is a general, (2)

and

If de Gaulle is a French general, then he is a general. (3)

All three sentences are true; only the last one, however, is what we would call logically true. Or compare the four closed sentences:

All men are mortal, (4)

All Americans are men, (5)

All Americans are mortal, (6)

and

If all men are mortal and all Americans are men, then all Americans are mortal. (7)

All four sentences are true; only the last one, however, is what we would call logically true.

The difference between (1)–(2), on one hand, and (3), on the other, can be explained as follows. (1) is true because the individual that 'de Gaulle' happens to designate is one of those that 'is French' happens to have as its extension; it could, however, be false if 'de Gaulle' designated another individual or 'is French' had another extension. Similarly, (2) is true because the individual that 'de Gaulle' happens to designate is one of those that 'is a general' happens to have as its extension; it could, however, be false if 'de Gaulle' designated another individual or 'is a general' had another extension. On the other hand, (3) turns out to be true no matter which individual 'de Gaulle' may designate or which sets of individuals 'is French' and 'is a general' may have as their respective extensions.

The difference between (4)–(6), on one hand, and (7), on the other, can be explained along the same lines. (4) is true because the individuals that 'is a man' happens to have as its extension are among those that 'is mortal' happens to have as its extension, (5) is true because the individuals that 'is an American' happens to have as its extension are among those that 'is a man' happens to have as its extension, and (6) is true because the individuals that 'is an American' happens to have as its extension are among those that 'is mortal' happens to have as its extension; (4), (5), and (6) could, however, be false if 'is a man,' 'is mortal,' and 'is an American' each had another extension. On the other hand, (7) turns out to be true no matter which sets of individuals 'is a man,' 'is mortal,' and 'is an American' may have as their respective extensions.

Generalizing the above, I first lay down the conditions under which a sentence (be it open or closed) of L is to be valid in L (D4.1); I then take a closed sentence of L to be logically true in L if it is valid in L (D4.2) and to be logically false in L if its negation is valid in L (D4.3). It can be shown, by means of a result of Kurt Gödel's, that a sentence P of L is valid in L if and only if every instance of P in L is true in L no matter which individuals (presumed for the occasion to be distinct from one another) the individual constants of L may designate in L, no matter which set of those individuals each one-place predicate of L may have as its extension in L, no matter which set of pairs of those individuals each two-place predicate of L may have as its extension in L, and so on; and, hence, that a closed sentence P of L is valid in L if and only if P is logically true in L in the sense just explained.[23] To condense matters, I write '$\vdash P$' for 'P is valid in L,' '$\vdash^{\infty} P$' for 'P is valid in L^{∞},' and '$\vdash^{N} P$' for 'P is valid in the sublanguage L^N of L^{∞}.'

D4.1. **(a)** $\vdash P \supset (Q \supset P)$;

(b) $\vdash (P \supset (Q \supset R)) \supset ((P \supset Q) \supset (P \supset R))$;

(c) $\vdash (\sim P \supset \sim Q) \supset (Q \supset P)$;

(d) $\vdash (\forall W)(P \supset Q) \supset (P \supset (\forall W)Q)$, *where the individual variable W of L is not free in P;*

(e) $\vdash (\forall W)P \supset P'$, *where P′ is like P except for containing free occurrences of an individual sign W′ of L at all the places where P contains free occurrences of the individual variable W of L;*

(f) *For each N from 1 on,* $\vdash^{N} P_1 \supset (P_2 \supset (\cdots \supset (P_N \supset (\forall W)P)\cdots))$, *where the individual variable W of* L^N *is free in P and where* P_i*, for each i from 1 to N, is like P except for containing occurrences of the i-th individual constant of* L^N *at all the places where P contains free occurrences of W;*

(g) $\vdash W = W$, *where W is an individual variable of L;*

(h) $\vdash W = W' \supset (P \supset P')$, *where P′ is like P except for containing free occurrences of the individual variable W′ of L at one or more places where P contains free occurrences of the individual variable W of L;*

(i) $\vdash \sim W = X$, *where W and X are two individual constants of L distinct from each other;*

(j) *If* $\vdash P$ *and* $\vdash P \supset Q$, *then* $\vdash Q$;

(k) *If* $\vdash P$, *then* $\vdash (\forall W)P$;

(l) *No sentence of L is valid in L unless its being so follows from (a)–(k).*

D4.2. *A closed sentence P of L is logically true in L if* $\vdash P$.

D4.3. *A closed sentence P of L is logically false in L if* $\vdash \sim P$.

The reader who has some familiarity with modern logic should recognize all the clauses of D4.1 except, possibly, (f) and (i). D4.1(i) mirrors the fact that any two distinct individual constants of L are expected to designate in L distinct individuals; D4.1(f) mirrors the fact that the individuals designated in L^N by the individual constants of L^N are expected to make up the entire universe of discourse of L^N. The individuals designated in L^∞ by the individual constants of L^∞ are likewise expected to make up the entire universe of discourse of L^∞. Since, however, the expressions of L^∞ are all presumed by D2.1 to be finite in length, this fact cannot be mirrored in D4.1.[24]

To illustrate the workings of D4.1, I prove of four sample bundles of sentences of L that they are valid in L. The reader is invited to prove on his own further consequences of D4.1–3 which I shall take for granted at various points in the book.

EXAMPLE 1: $\vdash (P \supset Q) \supset ((Q \supset R) \supset (P \supset R))$.

Proof: 1. By D4.1(b), D4.1(a), and D4.1(j),

$$\vdash (Q \supset R) \supset ((P \supset (Q \supset R)) \supset ((P \supset Q) \supset (P \supset R)));$$

hence by D4.1(b) and D4.1(j),

$$\vdash ((Q \supset R) \supset (P \supset (Q \supset R))) \supset ((Q \supset R) \supset ((P \supset Q) \supset (P \supset R)));$$

and hence by D4.1(a) and D4.1(j),

$$\vdash (Q \supset R) \supset ((P \supset Q) \supset (P \supset R)).$$

2. By a similar reasoning,

$$\vdash (((Q \supset R) \supset (P \supset Q)) \supset ((Q \supset R) \supset (P \supset R))) \supset (((P \supset Q) \supset ((Q \supset R) \supset (P \supset Q))) \supset ((P \supset Q) \supset ((Q \supset R) \supset (P \supset R)))).$$

3. By 1, D4.1(b), and D4.1(j),

$$\vdash ((Q \supset R) \supset (P \supset Q)) \supset ((Q \supset R) \supset (P \supset R)).$$

4. By 2, 3, and D4.1(j),

$$\vdash ((P \supset Q) \supset ((Q \supset R) \supset (P \supset Q))) \supset ((P \supset Q) \supset ((Q \supset R) \supset (P \supset R)));$$

and hence by D4.1(a) and D4.1(j),

$$\vdash (P \supset Q) \supset ((Q \supset R) \supset (P \supset R)).$$ [25]

EXAMPLE 2: $\vdash \sim\sim P \supset P$.

Proof: 1. By D4.1(a),

$$\vdash \sim P \supset (\sim Q \supset \sim P);$$

and hence by Example 1, D4.1(c), and D4.1(j),

$$\vdash \sim P \supset (P \supset Q).$$

2. By D4.1(a),

$$\vdash P \supset ((Q \supset P) \supset P);$$

hence by D4.1(b) and D4.1(j),

$$\vdash (P \supset (Q \supset P)) \supset (P \supset P);$$

and hence by D4.1(a) and D4.1(j),

$$\vdash P \supset P.$$

3. By 1,

$$\vdash \sim\sim P \supset (\sim P \supset \sim\sim\sim P).$$

But by D4.1(c),

$$\vdash (\sim P \supset \sim\sim\sim P) \supset (\sim\sim P \supset P).$$

Hence by Example 1 and D4.1(j),

$$\vdash \sim\sim P \supset (\sim\sim P \supset P);$$

hence by D4.1(b) and D4.1(j),

$$\vdash (\sim\sim P \supset \sim\sim P) \supset (\sim\sim P \supset P);$$

and hence by 2 and D4.1(j),

$$\vdash \sim\sim P \supset P.$$[26]

EXAMPLE 3 : $\vdash (\forall W)(P \supset Q) \supset ((\forall W)P \supset (\forall W)Q)$.

Proof: By D4.1(e),

$$\vdash (\forall W)(P \supset Q) \supset (P \supset Q).$$

But by Example 1, D4.1(e), and D4.1(j),

$$\vdash (P \supset Q) \supset ((\forall W)P \supset Q).$$

Hence by Example 1 and D4.1(j),

$$\vdash (\forall W)(P \supset Q) \supset ((\forall W)P \supset Q);$$

hence by D4.1(k),

$$\vdash (\forall W)((\forall W)(P \supset Q) \supset ((\forall W)P \supset Q));$$

hence by D4.1(d) and D4.1(j),

$$\vdash (\forall W)(P \supset Q) \supset (\forall W)((\forall W)P \supset Q);$$

and hence by Example 1, D4.1(d), and D4.1(j),

$$\vdash (\forall W)(P \supset Q) \supset ((\forall W)P \supset (\forall W)Q).$$

EXAMPLE 4: $\vdash W = X \supset (\forall Y)(W = Y \supset X = Y)$, *where* W *and* X *are two individual constants of* L.

Proof: Let W' and X' be two individual variables of L distinct from each other and from Y. By D4.1(h) and D4.1(k),

$$\vdash (\forall W')(\forall X')(\forall Y)(W' = X' \supset (W' = Y \supset X' = Y));$$

hence by D4.1(e) and D4.1(j),

$$\vdash (\forall Y)(W = X \supset (W = Y \supset X = Y));$$

and hence by D4.1(d), the hypothesis on W and X, and D4.1(j),

$$\vdash W = X \supset (\forall Y)(W = Y \supset X = Y).$$

The last two entries in this section deal with the notion of logical implication and logical equivalence. According to D4.4, a sentence P of L logically implies in L a sentence Q of L if $\vdash P \supset Q$; two sentences P_1 and P_2 of L logically imply in L a sentence Q of L if $\vdash P_1 \supset (P_2 \supset Q)$; three sentences P_1, P_2, and P_3 of L logically imply in L a sentence Q of L if $\vdash P_1 \supset (P_2 \supset (P_3 \supset Q))$; and so on. According to D4.5, a sentence P of L is logically equivalent in L to a sentence Q of L or, as I shall often put it, two sentences P and Q of L are logically equivalent in L if $\vdash P \equiv Q$.

D4.4. *Let* $P_1, P_2, \cdots, P_n$, *and* Q *be* $n + 1$ $(n \geq 1)$ *sentences of* L. *The sequence made up of* $P_1, P_2, \cdots$, *and* P_n *(in that order) logically implies* Q *in* L *if* $\vdash P_1 \supset (P_2 \supset (\cdots \supset (P_n \supset Q)\cdots))$.

D4.5. *A sentence* P *of* L *is logically equivalent in* L *to a sentence* Q *of* L *if* $\vdash P \equiv Q$.

The notions of logical implication and logical equivalence are vital to deductive logic. A conclusion, call it Q, is in fact deducible from n $(n \geq 1)$ premises, call them $P_1, P_2, \cdots$, and P_n, if (and only if) $P_1, P_2, \cdots$, and P_n logically imply Q; two sentences P and Q are deducible from each other, on the other hand, if (and only if) P and Q are logically equivalent. The reader will notice, by the way, that (6) on page 16 is logically implied by and hence deducible from (4) and (5) on the previous page. Our major concern here, however, is induction rather than deduction.

5. SET THEORY AND THE LANGUAGES L

Although the universe of discourse of L may, as is hinted in Section 3, consist partly or entirely of sets, we are in no position yet to talk in L about the various sets which draw their membership from that universe of discourse. I wish to remedy things here by grafting onto L an additional number of defined signs.

Needed first and foremost for talking in L about sets are sentences of the form:

W belongs to the set of all X's such that P,[27]

or, to use a current abbreviation,

$W \in \hat{X}(P)$,[28]

where W is an individual sign of L, X is an individual variable of L, and P is a sentence of L. Among them are to be found, for example,

$$\text{Peter} \in \hat{x}(\text{Likes}(x, \text{golf})), \tag{1}$$

$$w \in \hat{x}((\forall y)\text{Tries}(x,y)), \tag{2}$$

$$w \in \hat{x}((\forall w)(\text{Tries}(x,w) \supset \text{Likes}(x,w))), \tag{3}$$

and so on.

Since (1) is true if and only if Peter likes golf, we might think of defining $W \in (\hat{X}P)$ as P', where P' is like P except for containing occurrences of W at all the places where P contains free occurrences of X. The definition would suit some sentences, (1), for example, which might well pass as a rewrite of 'Likes(Peter,golf),' or (2), which might well pass as a rewrite of '$(\forall y)\text{Tries}(w,y)$.' It would unfortunately not suit others. (3), for example, might well pass as a rewrite of '$(\forall x)(\text{Tries}(w,x) \supset \text{Likes}(w,x))$,' or '$(\forall y)(\text{Tries}(w,y) \supset \text{Likes}(w,y))$,' or '$(\forall z)(\text{Tries}(w,z) \supset \text{Likes}(w,z))$,' and so on; it would definitely not pass, however, as a rewrite of '$(\forall w)(\text{Tries}(w, w) \supset \text{Likes}(w,w))$.'

A satisfactory alternative is, however, readily found. Let indeed $W \in \hat{X}(P)$ be defined as $(\exists X')(X' = W \,\&\, P')$, where X' is an individual variable of L both distinct from W and X and foreign to P and where P' is like P except for containing occurrences of X' at all the places where P contains free occurrences of X. The definition will suit (1), which may pass as a rewrite of '$(\exists w)(w =$ Peter & Likes$(w$, golf$))$,' for example, since '$(\exists w)(w =$ Peter & Likes$(w$, golf$))$' is logically equivalent to our previous 'Likes(Peter, golf).' It will suit (2), which may pass as a rewrite of '$(\exists z)(z = w$ & $(\forall y)$Tries$(z,y))$,' for example, since '$(\exists z)(z = w$ & $(\forall y)$Tries$(z,y))$' is logically equivalent to our previous '$(\forall y)$Tries(w,y).' Finally, it will suit (3), which may pass as a rewrite of '$(\exists y)(y = w$ & $(\forall w)($Tries$(y,w) \supset$ Likes$(y,w)))$,' for example, since '$(\exists y)(y = w$ & $(\forall w)($Tries$(y,w) \supset$ Likes$(y,w)))$' is logically equivalent to any one of our previous '$(\forall x)($Tries$(w,x) \supset$ Likes$(w,x))$,' '$(\forall y)($Tries$(w,y) \supset$ Likes$(w,y))$,' and '$(\forall z)($Tries$(w,z) \supset$ Likes$(w,z))$.'[29]

Also needed for talking in L about sets are sentences of the following forms:

W_1 and W_2 belong to the set of all X's such that P,

W_1, W_2, and W_3 belong to the set of all X's such that P,

and so on, which I condense to read:

$$W_1, W_2 \in \hat{X}(P),$$

$$W_1, W_2, W_3 \in \hat{X}(P),$$

and so on. They may respectively be treated as rewrites of

$$W_1 \in \hat{X}(P) \;\&\; W_2 \in \hat{X}(P),$$

$$W_1, W_2 \in \hat{X}(P) \;\&\; W_3 \in \hat{X}(P),$$

and so on.

D5.1. *Let X be an individual variable of L.*

(a) *$W \in \hat{X}(P)$ is defined as $(\exists X')(X' = W \;\&\; P')$, where X' is an individual variable of L both distinct from W and X and foreign to P and where P' is like P except for containing occurrences of X' at all the places where P contains free occurrences of X;*

(b) *$W_1, W_2, \cdots, W_{n+1} \in \hat{X}(P)$ is defined as*

$$W_1, W_2, \cdots, W_n \in \hat{X}(P) \;\&\; W_{n+1} \in \hat{X}(P),$$

where $n \geq 1$.

The letter '$\in$,' which I just grafted onto L, is known as the membership predicate; the expression $\hat{X}(P)$ is known as a set abstract; and the sentence P in $\hat{X}(P)$ is known as the defining condition of the set designated in L by $\hat{X}(P)$.[30] To condense matters, I adopt at this point four extra metalinguistic variables, the capitals 'A,' 'B,' 'C,' and 'D,' which are to range over the set abstracts of L. The selfsame letters, by the way, will be made in Chapter 2 to range over sets rather than set abstracts.

Defined next are three set predicates, '$=$,' '$\subset$,' and '$\emptyset$.' $A = B$ may be read: A is identical with B, A is the same as B, or A is B; $A \subset B$ may be read: A is a subset of B or A is included in B; $A \emptyset B$, finally, may be read: A does not overlap B, A and B are disjoint, or A and B are mutually exclusive.

D5.2. *$A = B$ is defined as $(\forall W)(W \in A \equiv W \in B)$, where W is an individual variable of L foreign to A and B.*

D5.3. *$A \subset B$ is defined as $(\forall W)(W \in A \supset W \in B)$, where W is as in D5.2.*

D5.4. *$A \emptyset B$ is defined as $\sim (\exists W)(W \in A \,\&\, W \in B)$, where W is as in D5.2.*

According to D5.2, two sets are identical with each other when they have the same members; according to D5.3, one set is a subset of another when all the members of the first are members of the second; and, according to D5.4, two sets do not overlap each other when they have no member in common. To use people as our individuals, the set of all American Presidents is identical with the set of all American Commanders-in-chief; the set of all American Presidents up to 1960 is a subset of the set of all Protestants; and the set of all Republican nominees for President up to 1960 does not overlap the set of all Catholics. Or, to use events as our individuals, the set of all throws with two dice resulting in sum six is identical with the set of all those resulting in (1, 5), (2, 4), (3, 3), (4, 2), or (5, 1); the set of all throws resulting in sum four or sum six is a subset of the set of all those resulting in an even sum; and the set of all throws resulting in an even sum does not overlap the set of all those resulting in (1, 2), (1, 4), or (1, 6).

As the reader undoubtedly knows, a set of size n ($n \geq 0$) has 2^n subsets. The so-called null set has one subset: itself; a set consisting of one member, say Washington, has two subsets: the null set and the set consisting of Washington; a set consisting of two members, say Washington and Jefferson, has four subsets: the null set, the set consisting of Washington, the set consisting of Jefferson, and the set consisting of Washington and Jefferson; and so on. Likewise, a set of size $\aleph_0$ has $2^{\aleph_0}$ subsets, a set of size $2^{\aleph_0}$ has $2^{2^{\aleph_0}}$ subsets, and so on. That a set of size n ($n \geq 0$) has 2^n subsets can be proved, by the way, in the languages L so long as the set is given by enumeration.[31]

Defined next are three set functors: '$-$,' which helps to designate the so-called complement (or logical difference) of a set, '$\cap$,' which helps to designate the so-called intersection (or logical product) of two sets, and '$\cup$,' which helps to designate the so-called union (or logical sum) of two sets.

D5.5. $\overline{A}$ *is defined as* $\hat{W} \sim (W \in A)$, *where* W *is an individual variable of* L *foreign to* A.

D5.6. $(A) \cap (B)$ *is defined as* $\hat{W}(W \in A \;\&\; W \in B)$, *where* W *is an individual variable of* L *foreign to* A *and* B.

D5.7. $(A) \cup (B)$ *is defined as* $\hat{W}(W \in A \vee W \in B)$, *where* W *is as in D5.6.*

According to D5.5, $\overline{A}$ designates the set, known as the complement of the set designated by A, of all the individuals which do not belong to the set designated by A; according to D5.6, $A \cap B$ designates the set, known as the intersection of the sets designated by A and B, of all the individuals which belong to each one of the sets designated by A and B; and, according to D5.7, $A \cup B$ designates the set, known as the union of the sets designated by A and B, of all the individuals which belong to at least one of the sets designated by A and B. To use people as our individuals, the set of all minors is the complement of the set of all the people who have come of age; the set of all teen-age delinquents is the intersection of the set of all teen-agers and the set of all delinquents; and the set of all college upperclassmen is the union of the set of all college juniors and the set of all college seniors. Or, to use events as our individuals, the set of all the hands in which a poker player is not dealt an ace is the complement of the set of all those in which he is dealt at least one; the set of all the hands in which he is dealt both red aces is the intersection of the set of all those in which he is dealt the ace of diamonds and the set of all those in which he is dealt the ace of hearts; and the set of all the hands in which he is dealt at least three aces is the union of the set of all those in which he is dealt three aces and the set of all those in which he is dealt all four aces.

Defined next is a fourth set predicate, 'P,' with P(A,B,C) meant to be read: A and B constitute a partition of C.

D5.8. P(A,B,C) *is defined as* $(A \subset C \;\&\; B \subset C) \;\&\; (A \not o B \;\&\; A \cup B = C)$.

According to D5.8, two sets, called for the occasion cells, constitute a partition of a third set when the two sets are subsets of the third, do not overlap each other, and add up to the third. To use people as our individuals, the set of all males and the set of all females constitute a partition of the set of all human beings; to use events as our individuals, the set of all the tosses in which a given coin lands heads up and the set of all those in which it lands tails up constitute a partition of the set of all the tosses of the coin. D5.8 is easily extended to cover partitions into three or more cells; I leave the matter to the reader.

Defined next are two set abstracts, one designating the so-called universal set $\vee$, the other the so-called null set $\wedge$.

D5.9. '$\vee$' *is defined as* '$\hat{w}(w = w)$.'

D5.10. '$\wedge$' *is defined as* '$\overline{\vee}$.'

Since every instance of '$w = w$' in L^∞ is true in L^∞, '$\vee$' designates in L^∞ what I have called above the universe of discourse of L^∞ and referred to as $\vee^1_{L^\infty}$. Similarly, since every instance of '$w = w$' in L^N is true in L^N, '$\vee$' designates in L^N what I have called above the universe of discourse of L^N and referred to as $\vee^1_{L^N}$. $\vee$ is of size $\aleph_0$ in L^∞ and of size N in L^N; $\wedge$, on the other hand, is of size 0 in every language L. As the reader may have noticed, the sentence P in a set abstract $\hat{X}P$ may be closed as well as open. It can be shown in the former case that the set designated in L by $\hat{X}P$ is identical with $\vee$

when P is true in L, identical with $\wedge$ when P is false in L. I shall avail myself of this result in Chapter 3.

Defined next are abstracts designating sets given, as the phrase goes, by listing or enumeration.

D5.11. **(a)** *$\{W\}$ is defined as $\hat{X}(X = W)$, where X is an individual variable of L distinct from W;*

(b) *$\{W_1, W_2, \cdots, W_{n+1}\}$ is defined as $\{W_1, W_2, \cdots, W_n\} \cup \{W_{n+1}\}$, where $n \geq 1$.*

According to D5.11(a), $\{W\}$, where W is an individual constant of L, designates the set whose only member is the individual designated by W; according to D5.11(b), $\{W_1, W_2, \cdots, W_{n+1}\}$, where W_1, W_2, $\cdots$, and W_{n+1} are $n + 1$ individual constants of L, designates the set whose only members are the individuals designated by W_1, W_2, $\cdots$, and W_{n+1}. The former set is of size 1 and, for that reason, is often called a unit set. The latter is of size $n + 1$ when W_1, W_2, $\cdots$, and W_{n+1} are all distinct from one another; otherwise, it is of size m, where m is the number of distinct individual constants among W_1, W_2, $\cdots$, and W_{n+1}. {Caesar, Crassus, Pompey}, for example, is of size 3, {Caesar, Caesar, Pompey} of size 2, and {Caesar, Caesar, Caesar} of size 1.

The sets designated by $\{W_1, W_2, \cdots, W_n\}$ have been called non-ordered sequences. They are a breed apart from the sequences I talked about in Section 2, sequences often designated by $\langle W_1, W_2, \cdots, W_n\rangle$. Whereas, for example, {Caesar, Pompey} and {Pompey, Caesar} are the same set under different guises, ⟨Caesar, Pompey⟩ and ⟨Pompey, Caesar⟩ are two different sequences, one being made up of the two individuals Caesar and Pompey in that order, the other of the two individuals Pompey and Caesar in that order. Whereas also {Caesar, Caesar} and {Caesar} are the same set (under different guises again) and hence are both of size 1, ⟨Caesar, Caesar⟩ and ⟨Caesar⟩ are two different sequences, one being made up of two individuals, the other of one individual. Thus the order in which two or more individuals turn up in a given sequence is important, as is the number of times a given individual turns up in the sequence.[32]

Some of the contexts in which '⟨' and '⟩' are commonly used can be defined in L; others, however, cannot, unless additions are made to the primitive vocabulary of L. The limitation is a severe one. It does not affect, fortunately, the so-called metalanguage ML in which I have talked and shall keep talking about L.[33] I have already mentioned sequences of members of V_L^1, when talking about L, and shall do so again; I have also mentioned sequences of signs of L and shall do so again, in Chapter 3, especially, where I deal at length with sequences of individual constants of L^N.

Defined last is a numerical functor, the functor 'S,' to be read 'The size of,' as it occurs in the four contexts $\mathsf{S}(A) = n$, $\mathsf{S}(A) = m + n$, $\mathsf{S}(A) = m \cdot n$, and $\mathsf{S}(A) = m^n$. I avail myself in D5.12–15 of the canonical notation '0′' for '1,' '0″' for '2,' '0‴' for '3,' and so on, and presume, as a result, that the metalinguistic variables 'm' and 'n' range over such expressions as '0,' '0′,' '0″,' and so on.

D5.12. (a) $\mathsf{S}(A) = 0$ *is defined as* $A = \wedge$;

(b) $\mathsf{S}(A) = n'$ *is defined as* $(\exists W)(W \in A \mathbin{\&} \mathsf{S}(A \cap \overline{\{W\}}) = n)$, *where* W *is an individual variable of L foreign to* A.

According to D5.12, $\mathsf{S}(A) = 0'$ (i.e., $\mathsf{S}(A) = 1$) is a rewrite in L of $(\exists W)(W \in A \mathbin{\&} \mathsf{S}(A \cap \overline{\{W\}}) = 0)$ and hence of $(\exists W)(W \in A \mathbin{\&} A \cap \overline{\{W\}} = \wedge)$, where W is foreign to A in both cases; $\mathsf{S}(A) = 0''$ (i.e., $\mathsf{S}(A) = 2$) is a rewrite in L of $(\exists W)(W \in A \mathbin{\&} \mathsf{S}(A \cap \overline{\{W\}}) = 0')$, hence of $(\exists W)(W \in A \mathbin{\&} (\exists X)(X \in A \cap \overline{\{W\}} \mathbin{\&} \mathsf{S}((A \cap \overline{\{W\}}) \cap \overline{\{X\}}) = 0))$, and hence of $(\exists W)(W \in A \mathbin{\&} (\exists X)(X \in A \cap \overline{\{W\}} \mathbin{\&} (A \cap \overline{\{W\}}) \cap \overline{\{X\}} = \wedge))$, where W is foreign to A in all three cases and X is foreign to A and distinct from W in the last two; and so on.

D5.13. (a) $\mathsf{S}(A) = m + 0$ *is defined as* $\mathsf{S}(A) = m$;

(b) $\mathsf{S}(A) = m + n'$ *is defined as* $\mathsf{S}(A) = (m + n)'$.

D5.14. (a) $\mathsf{S}(A) = m \cdot 0$ *is defined as* $\mathsf{S}(A) = 0$;

(b) $\mathsf{S}(A) = m \cdot n'$ *is defined as* $\mathsf{S}(A) = m + (m \cdot n)$.

D5.15. (a) $\mathsf{S}(A) = m^0$ *is defined as* $\mathsf{S}(A) = 0'$;

(b) $\mathsf{S}(A) = m^{n'}$ *is defined as* $\mathsf{S}(A) = m \cdot m^n$.

According to D5.15, $\mathsf{S}(A) = 0''^{0''}$ (i.e., $\mathsf{S}(A) = 2^2$) is a rewrite in L of $\mathsf{S}(A) = 0'' \cdot (0'' \cdot 0')$; hence, according to D5.14, a rewrite in L of

$\mathsf{S}(A) = 0''\cdot(0'' + 0)$; hence, according to D5.13, a rewrite in L of $\mathsf{S}(A) = 0''\cdot 0''$; hence, according to D5.14, a rewrite in L of $\mathsf{S}(A) = 0'' + (0'' + 0)$; hence, according to D5.13, a rewrite in L of $\mathsf{S}(A) = 0''''$ (i.e., $\mathsf{S}(A) = 4$).[34]

The same four contexts with '$\leq$,' '$<$,' '$\geq$,' and '$>$' in place of '$=$' are definable in all the languages L, as the reader may verify. The five contexts $\mathsf{S}(A) = \aleph_0$, $\mathsf{S}(A) \leq \aleph_0$, $\mathsf{S}(A) < \aleph_0$, $\mathsf{S}(A) \geq \aleph_0$, and $\mathsf{S}(A) > \aleph_0$ are definable in the various sublanguages of L^∞.[35] The two contexts $\mathsf{S}(A) \leq \aleph_0$ and $\mathsf{S}(A) > \aleph_0$ are definable in L^∞.[36] The three contexts $\mathsf{S}(A) = \aleph_0$, $\mathsf{S}(A) < \aleph_0$, and $\mathsf{S}(A) \geq \aleph_0$, on the other hand, cannot be defined in L^∞ unless additions are again made to the primitive vocabulary of L^∞.

Besides talking of the size of a set, we also talk of the relative frequency (**Rf**, for short) with which the members of one set, often called a reference set, belong to another set. Few of the contexts in which the numerical functor '**Rf**' turns up are definable in L unless additions are again made to the primitive vocabulary of L. Relative frequencies play such a part, however, in the following chapters as to rate a word of explanation here.

By the relative frequency with which the members of a finite set belong to another set we understand the ratio of the size of the intersection of the two sets to the size of the first. The relative frequency with which the members of a given audience are male is thus the ratio of the number of people in the audience who are male to the number of people in the audience or, if you wish, the proportion of males in the audience. By the relative frequency with which the members of a denumerably infinite and serially ordered set belong to another set we understand the limit (if any) for increasing i of the relative frequency with which the first i members of the first set belong to the second.[37] The relative frequency with which the natural numbers from 1 on, as arranged in the serial order 1, 2, 3, 4, 5, 6, and so on ad infinitum, are odd is thus the limit ($\frac{1}{2}$, as it happens) for increasing i of the relative frequency with which the first i numbers in question are odd. The reader should note, by the way, that the members of a denumerably infinite set can be arranged in many different serial orders and that the limit for increasing i of the relative frequency with which the first i of them belong to another set often varies with the serial order in which the members of the set are arranged.[38]

NOTES

[1] My languages L are patterned after R. Carnap's languages $\mathfrak{L}$ in *Logical Foundations of Probability*, §§ 14–20. Many of the definitions laid down in this chapter already appear in Carnap; others are new. For a fuller treatment of the logic of sentences, see the author's *An Introduction to Deductive Logic*, W. V. O. Quine's *Methods of Logic*, A. Church's *Introduction to Mathematical Logic*, and references supplied therein; for a fuller treatment of the logic of sets, see P. Suppes' *Axiomatic Set Theory*, A. A. Fraenkel's *Abstract Set Theory*, W. V. O. Quine's *Mathematical Logic*, and references supplied therein.

[2] A set is said to be denumerably or, sometimes, countably infinite when it has as many members as there are natural numbers or unsigned integers; on this point, see Fraenkel, *loc. cit.*, § 3.

[3] To condense matters, I shall frequently omit the qualification "in alphabetical order" when talking of the i-th individual sign—in alphabetical order—of a given language L.

[4] More specifically, closed sentences in the indicative mood.

[5] That is, 'Peter knows everything,' 'All men are mortal,' and 'Everything is self-identical.'

[6] Ad hoc expressions like 'Individual No. 1,' 'Individual No. 2,' and so on, can always be drafted to serve as the individual constants of L^∞. There is accordingly nothing unrealistic in presuming that L^∞ is fitted with infinitely many individual constants.

[7] By the size of a universe of discourse and, more generally, of a set, I understand, of course, the number of members of the universe or of the set in question (see D5.12). Languages with larger universes of discourse than the above can be and have been devised; they raise delicate problems, however, when it comes to allotting statistical and inductive probabilities to their sentences.

[8] To condense matters, I shall hereafter write 'L' for 'the languages L' or, as the context may suggest, 'a language L.'

[9] In much of the literature (publications of mine included) expressions (in the sense of D2.1) are called formulas, sentences (in the sense of D2.2) called well-formed formulas, closed sentences (in the sense of D2.4) called sentences, and open sentences (in the sense of D2.5) called quasi-sentences, matrices, propositional functions, sentential functions, and so on; closed sentences are also called statements and, by the same token, open sentences called quasi-statements.

[10] The n individual signs W_1, W_2, $\cdots$, and W_n in D2.2(a) need not be distinct from one another; the same holds true of the two individual signs W and X in

D2.2(b), of the two sentences P and Q in D2.2(d), and, unless otherwise indicated, of all the signs or sequences of signs of L mentioned in a definition or a theorem.

[11] The subsequence $(\forall W)(Q)$ of P may be P itself; see example (2) on page 7.

[12] For proof that $\aleph_0{}^n$ equals $\aleph_0$, see Fraenkel, *loc. cit.*, §§ 6–7.

[13] For further remarks on D2.7–10, see Note 20 of this chapter.

[14] D2.11 is a definition of a special sort. To retrieve the original of '$\mathsf{D}(w,x,y)$,' for example, we must first consult D2.11(b), which shows '$\mathsf{D}(w,x,y)$' to be a rewrite in L of '$\mathsf{D}(w,x)$ & $(\mathsf{D}(y,w)$ & $\mathsf{D}(y,x))$,' then consult D2.11(a), which shows '$\mathsf{D}(w,x)$ & $(\mathsf{D}(y,w)$ & $\mathsf{D}(y,x))$' to be a rewrite in L of '$\sim w = x$ & $(\sim y = w$ & $\sim y = x)$.' Definitions of this kind are often called recursive definitions.

[15] In the informal remarks which opened this section I referred to sequences made up of one individual as individuals, to sequences made up of two individuals as pairs of individuals, and to sequences made up of three individuals as triples of individuals. Note that the sequences in question need not be made up of distinct individuals; the same remark applies to the sequences in D3.1(d).

[16] One set is said to be a subset of another when all the members of the first set are members of the second (see D5.3).

[17] The null set is the set to which nothing belongs (see D5.10); it is a subset of each one of $\mathsf{V}^1_{L\infty}$, $\mathsf{V}^2_{L\infty}$, $\cdots$, and $\mathsf{V}^n_{L\infty}$.

[18] It is less common, of course, to give the extension of an n-place predicate G by listing or enumeration than by description. In the former case, D3.1 and two further definitions, D3.2–3, suffice to tell whether a closed sentence $G(W_1, W_2, \cdots, W_n)$ of L is true in L or false in L; in the latter, observation, experimentation, and the like may be needed as well.

[19] Or, in view of D3.1(a) and D3.1(c), *If W is an individual constant of L, then $W = W$ is true in L.*

[20] It follows from D2.7–10 and D3.2(c)–(e) that: (1) a closed sentence P & Q is true in L if and only if P is true in L and Q is true in L, (2) a closed sentence $P \vee Q$ is true in L if and only if P is true in L and/or Q is true in L, (3) a closed sentence $P \equiv Q$ is true in L if and only if P is true in L if Q is true in L and vice versa, and (4) a closed sentence $(\exists W)P$ is true in L if and only if some instance of P in L is true in L. These four consequences should vindicate D2.7–10.

[21] Definition D3.2 is both formally and materially adequate in the sense of Tarski. For further details on this matter, see A. Tarski, "The Semantic Conception of Truth and the Foundations of Semantics."

[22] Duplications will occur among the individual constants X_1, X_2, $\cdots$, and X_n if the members of V^1_L assigned by **Asst** to W_1, W_2, $\cdots$, and W_n are not all distinct from one another. Another notion of satisfaction is also current, which does not require the notion of truth to be already on hand and by means of which

the notion of truth is frequently defined. For further details, see Tarski, *loc. cit.*

[23] In much of the literature a sentence P of L would be said to be a theorem in L if P is as in D4.1, and to be valid in L if every instance of P in L is true in L no matter which individuals etc. Gödel's result would then be used to show that a sentence P of L is a theorem in L if and only if P is valid in L. Abridging things, I dispense here with the notion of theoremhood in L, call valid in L such sentences of L as would normally be deemed theorems in L, and use Gödel's result to claim that a sentence P of L is valid in L if and only if every instance of P in L is true in L no matter which individuals etc. For an exposition of Gödel's result, see the author's *An Introduction to Deductive Logic*, pp. 209–218.

[24] Nor, by the way, does it have to be mirrored in L. This oddity, a variant of the so-called Skolem paradox, will be examined in *Truth and Estimated Truth*; it was brought to my attention by Professor Kanger.

[25] Part of the proof was suggested to me by Miss Nancy Spencer, a former student of mine at Bryn Mawr College.

[26] Note that by virtue of Example 2 and D2.8 $\sim P \vee P$ is also valid in L. One way of showing that a sentence P of L in which defined signs of L turn up is valid in L, is to retrieve the original, say P', of P, show that P' is valid in L, and then appeal to whichever definition or definitions will convert P' into P.

[27] Equivalently, W is a member of the set of all X's such that P, W is one of the X's such that P, and so on.

[28] Another current abbreviation is $W \in \{X \mid P\}$.

[29] The equivalences in question are special cases of the following theorem: *Let W' be distinct from W and foreign to P and let P' be like P except for containing occurrences of W' at all the places where P contains free occurrences of W. Then* $\vdash (\exists W')(W' = W \,\&\, P') \equiv P$.

[30] The sets which D5.1 enables us to talk about, though made up of members of the universe of discourse of L, do not belong to that universe of discourse and, hence, can only be said to be *virtually* (as opposed to *really*) designated in L by the set abstracts of L; the qualification, however, does not affect our purpose. Note, by the way, that the set designated in L by $\hat{W}G(W)$, where G is a one-place predicate of L, is what I called above the extension of G in L.

[31] Proof that a set of size n has 2^n subsets often runs as follows. It is first shown that the set has $\binom{n}{i}$ subsets of size i ($0 \leq i \leq n$), where $\binom{n}{i}$ is defined as $n!/i!(n-i)!$. It is then shown (this is Newton's Binomial Theorem) that

$$\sum_{i=0}^{n} \binom{n}{i} r_1{}^i r_2{}^{n-i} = (r_1 + r_2)^n,$$

where r_1 and r_2 are any two real numbers. With r_1 and r_2 both taken to be 1, it then follows that the set has 2^n subsets. Proof that a set of size n has $\binom{n}{i}$ subsets of size i will be sketched in Note 38 of Chapter 2. Proof of the Binomial

Theorem may be found, for example, in J. Neyman's *First Course in Probability and Statistics*, pp. 39–43.

32 For a proof that m ($m \geq 1$) individuals can be arranged in m^2 pairs, m^3 triples, and so on, see Note 38 of Chapter 2. The proof is analogous to the one I used in Section 2 to show that a sentence of L^N in which n ($n \geq 1$) individual variables of L^N are free has N^n instances in L^N.

33 The same holds true of other set-theoretic signs which cannot be grafted onto L, but are presumed to be part and parcel of ML.

34 D5.13–15 are patterned after celebrated definitions of '$m + n$,' '$m \cdot n$,' and 'm^n' in the arithmetic of natural numbers.

35 The first, fourth, and fifth contexts may be treated as rewrites in L^N of any logical falsehood of L^N, the second and third as rewrites in L^N of any logical truth of L^N. The definitions are suitable only because the universe of discourse of L^N is, by stipulation, of size N.

36 The first context may be treated as a rewrite in L^∞ of any logical truth of L^∞, the second as a rewrite in L^∞ of any logical falsehood of L^∞. The definitions are suitable only because the universe of discourse of L^∞ is, by stipulation, denumerably infinite in size.

37 A denumerably infinite set is said to be serially ordered when, informally speaking, each one of its members is identifiable as the first member, or the second member, or the third member, and so on, of the set, or, more formally speaking, when a function is supplied which maps the said set one-to-one onto the set of all the natural numbers from 1 on. Serially ordered sets go by many other names; among them are: 'linearly ordered sets,' 'simply ordered sets,' and 'strictly simply ordered sets.' On this whole matter, see R. L. Wilder, *Introduction to the Foundations of Mathematics*, pp. 44–46 and 97–98, and Suppes, *loc. cit.*, Chapter 3.

38 For example, when the natural numbers from 1 on are arranged in the serial order 1, 2, 4, 3, 6, 8, 5, 10, 12, and so on ad infinitum, the limit for increasing i of the relative frequency with which the first i of them are odd turns out to be $\frac{1}{3}$ rather than $\frac{1}{2}$.

2 STATISTICAL PROBABILITIES: PART ONE

In this second chapter I deal with statistical probabilities as measurements on sets. First, I show how sets are currently allotted absolute probabilities and pairs of sets conditional ones (Section 6); discuss and illustrate in some detail what is meant by the statistical probability of a set or of a pair of sets (Section 7); and go over some particulars of Section 6 which may be new to the reader, among them the allotting of unequal weights to the members of a probability set (Section 8). Then, I take up so-called random functions and show how they are currently allotted statistical probabilities (Section 9); discuss in connection with random sampling two celebrated distributions, the hypergeometric one and the binomial one (Section 10); and pass on to the problem of estimating statistical probabilities by means of sample soundings (Section 11).

Since sets (as opposed to sentences) are allotted only one kind of probability, I shall write 'probability' for 'statistical probability' in much of the chapter.

6. STATISTICAL PROBABILITIES ALLOTTED TO SETS

The various sets which draw their membership from a given finite set, dubbed for the occasion a probability set and referred to as *PS*,[1] may be allotted probabilities as follows. According to one procedure real numbers, called weights, are first allotted to the various members of *PS*, the numbers in question being subject to the following two restrictions:
(a1) each one of them must be non-negative, and
(a2) their sum must be equal to 1.[2]
A number, called an absolute probability or—for short—a probability, is then allotted to each subset *A* of *PS* (and hence to *PS*),[3] the number in question being the combined weights of the members of *PS* which belong to *A*.[4]

According to a second procedure real numbers, called absolute probabilities or—for short—probabilities, are directly allotted to the various subsets of *PS*, the numbers in question being subject to the following three restrictions:
(b1) each one of them must be non-negative,
(b2) the number allotted to *PS* must be 1, and
(b3) the number allotted to the union $A \cup B$ of two non-overlapping subsets *A* and *B* of *PS* must be equal to the sum of the numbers respectively allotted to *A* and *B*.
To round out the picture, a number, called a weight, may then be allotted to each member of *PS*, the number in question being the probability of the unit set of the said member of *PS*.[5]

Since any number which passes as a weight by either procedure does so by the other and any number which passes as a probability by either procedure does so again by the other, the two procedures are easily seen to yield the same allotments and hence can be used interchangeably.

One allotment of weights and probabilities stands out: the equiprobable, or finite-frequency, allotment, or (as I shall take the liberty of calling it here) relative frequency allotment. Under this allotment:

(1) the weight of a member of *PS* is equal to $1/n$, n being the size of *PS*, and hence equal to the relative frequency with which the members of *PS* belong to the unit set of the member in question, and

(2) the probability of a subset A of *PS* is equal to m/n, m being the size of A and n the size of *PS*, and hence equal to the relative frequency with which the members of *PS* belong to A.[6]

Note, however, that $2^{\aleph_0}$ allotments are endorsed here altogether and that some of them are occasionally preferred to the relative frequency allotment.

Proof that the allotments endorsed here are $2^{\aleph_0}$ in number is as follows. Let *PS* be of size n ($n \geq 2$) and let its n members be arranged —for the sake of argument— in a specified order. A weight may be allotted to the first member of *PS* in as many ways as there are real numbers in the interval [0, 1] which do not exceed 1, that is, in $2^{\aleph_0}$ ways; a weight may next be allotted to the second member of *PS* in as many ways as there are real numbers in the interval [0, 1] which, once added to the first weight, do not exceed 1, that is, in one way if the first weight was equal to 1, in $2^{\aleph_0}$ ways if it was not; a weight may next be allotted to the third member of *PS* in as many ways as there are real numbers in the interval [0, 1] which, once added to the first two weights, do not exceed 1, that is, in one way if the first two weights added up to 1, in $2^{\aleph_0}$ ways if the first two weights did not; $\cdots$; and a weight may finally be allotted to the n-th member of *PS* in as many ways as there are real numbers in the interval [0, 1] which, once added to the first $n - 1$ weights, equal 1, that is, in one way. But the result of multiplying $2^{\aleph_0}$ either by 1 or by itself $n - 1$ times is equal to $2^{\aleph_0}$.[7] The number of allotments endorsed here thus proves to be $2^{\aleph_0}$.

The sets which draw their membership from *PS* may be allotted probabilities not only one by one but also two by two. This is usually done as follows. Let absolute probabilities be allotted as above to the various subsets of *PS*; let two subsets A and B of *PS* be on hand; and let the absolute probability of B (a subset often known as a reference set) be non-zero. A number, called the conditional probability of A given B, is then allotted to the pair of subsets, the number in question being the ratio of the absolute probability of the intersection $A \cap B$ of A and B to the absolute probability of B.

One allotment of conditional probabilities also stands out, the relative frequency allotment. Under this allotment the conditional probability of two subsets A and B of PS is equal to m/n, m being the size of $A \cap B$ and n the size of B, and hence equal to the relative frequency with which the members of B belong to A. Note, however, that $2^{\aleph_0}$ allotments are again endorsed here and that some of them are occasionally preferred to the relative frequency allotment.

Conditional probabilities, I may point out, meet the following four requirements:

(c1) the conditional probability of a set A given a set B is non-negative,

(c2) the conditional probability of a set A given that set A is equal to 1,

(c3) the conditional probability of the intersection $A \cap B$ of two sets A and B given a set C is equal to the conditional probability of A given the intersection $B \cap C$ of B and C times the conditional probability of B given C, and

(c4) the conditional probability of the complement $\overline{A}$ of a set A given a set B is equal to 1 minus the conditional probability of A given B.

Analogues of (c1)–(c4) will turn up in Chapter 4 as requirements to be met by inductive probabilities.

PS, our probability set, was presumed above to be finite in size. The subsets of a denumerably infinite PS may be allotted absolute probabilities by suitable variants of the two procedures of page 34.[8] When PS is serially ordered, they may also be allotted absolute probabilities by a limit procedure. Let every member of PS be identifiable as the first member, or the second member, or the third member, and so on, of PS; let PS_i, for each i from 1 on, consist of the first i members of PS; and let the members of PS_i, for each i from 1 on, be allotted weights by either one of the two procedures of page 34. A subset A of PS may then be allotted as its absolute probability the limit (if any) for increasing i of the combined weights of the members of PS_i which belong to A.[9] Once the subsets of a denumerably infinite PS have been allotted absolute probabilities, pairs of those subsets may finally be allotted conditional probabilities by the procedure of pages 35–36.

One allotment of conditional probabilities stands out again. Assume that PS is serially ordered and that, for each i from 1 on, the weight of a member of PS_i is taken to be $1/i$. By the limit procedure I described in the last paragraph, the absolute probability of a subset A of PS will prove to be the limit (if any) for increasing i of the relative frequency with which the members of PS_i belong to A, and the conditional probability of two subsets A and B of PS prove to be the limit (if any) for increasing i of the relative frequency with which the members of the intersection $PS_i \cap B$ of PS_i and B belong to A.

Some accounts of the allotment run more simply. Given any two sets A and B, the second of which is presumed to be denumerably infinite and serially ordered, the conditional probability of A given B is sometimes taken to be the limit (if any) for increasing i of the relative frequency with which the first i members of B belong to A.[10] My account, however, may be more in line with recent trends in probability theory, where conditional probabilities are allotted to pairs of subsets of a preassigned probability set PS.

The allotment bears some likeness to the one championed by Richard von Mises, Hans Reichenbach, and others,[11] and will be called here—as in much of the literature—the relative frequency allotment.[12] Von Mises, for one, would further require that the members of B occur at random in A if the limit for increasing i of the relative frequency with which the first i members of B belong to A is to qualify as the conditional probability of A given B. The notion of random occurrence has proved, however, difficult to define, and von Mises' restriction, as a result, is frequently waived.[13]

Subsets and pairs of subsets of a non-denumerably infinite PS may finally be allotted probabilities by related, though more complicated procedures, which, for lack of space, I cannot recount here.[14]

Probabilities, so far, have been allotted to the subsets of a probability set *PS* and hence to sets of members of *PS*. They may also be allotted to sets of pairs, triples, and so on, of members of *PS*. When *PS* is finite, weights, subject to restrictions (a1)–(a2) on page 34, are first allotted to the pairs of members of *PS*, the triples of members of *PS*, and so on;[15] the absolute probability of a set of pairs, or triples, and so on, of members of *PS* is then reckoned as on page 34, that is, as the combined weights of the pairs, or triples, and so on, which belong to the set.[16] When *PS* is denumerably infinite and serially ordered, sets of pairs, triples, and so on, of its members may be allotted absolute probabilities by a limit procedure as on pages 36–37. The whole matter will prove of interest in Chapter 3, where I allot statistical probabilities to sentences of *L* containing free occurrences of two or more individual variables of *L*.

I showed in Section 5 how set abstracts of a sort can be grafted onto *L*. With the above instructions on hand the reader may go back to Chapter 1 and, after making such additions to D1.1–2 as may be needed to reproduce in *L* the mathematics of the preceding pages, allot probabilities to the subsets (and pairs of subsets) of $\vee$. I shall not do so myself, my concern here being more with sets in general than with those mentioned in *L*. In Chapter 3, however, I shall allot to the defining conditions of the subsets of $\vee$ probabilities which answer the probabilities allotted here to sets. My task, by the way, will prove to be lighter than the reader's. Whereas indeed the subsets of $\vee$ must be allotted probabilities in *L* itself, the sentences of *L* may be allotted probabilities in the metalanguage *ML* of *L* rather than in *L* and, hence, without any addition being made to D1.1–2.

7. COMMENTS AND ILLUSTRATIONS

The reader, as he fought his way through Section 6, may have wished for some illustrations. I purposely refrained, however, from supplying any until I had first discussed what can be meant, at a presystematic level, by the probability of a set.

Jerzy Neyman, whom I shall follow on this point, suggests that the probability of a subset *A* of a probability set *PS* be thought of as the probability of a member of *PS* belonging to *A*.[17] The expression 'a member of *PS*' is elliptic, to be sure, and liable as a result to cause misunderstandings. Paraphrases of it abound, however, in the literature, namely, 'an unspecified member of *PS*,' 'an arbitrary member of *PS*,' 'an arbitrarily chosen member of *PS*,' 'a randomly chosen member of *PS*,' and so on. Availing myself here of the first one,[18] I shall thus think of

(1) the weight of a member, say *a*, of *PS* as the probability (in a presystematic sense of the word) of an unspecified member of *PS* belonging to $\{a\}$ and hence being *a*,[19]

(2) the probability of a subset, say *A*, of *PS* as the probability (in a presystematic sense of the word) of an unspecified member of *PS* belonging to *A*, and

(3) the probability of one subset, say *A*, of *PS* given another subset, say *B*, of *PS* as the probability (in a presystematic sense of the word) of an unspecified member of *PS* belonging to *A* given that the member of *PS* in question belongs to *B*.

That membership in a subset *A* of *PS*, for one thing, and membership of an unspecified member of *PS*, for another, are at issue when we talk of the probability of *A*, may be clear from Section 6, where I reckoned the probability in question by adding the weights of the members of *PS* which belong to *A*, thus bypassing, on one hand, the members of *PS* which do not belong to *A* and withholding, on the other, the identity of those which do. That membership of an unspecified member of *PS* is at issue should also be clear from Chapter 3, where the probability of a subset (other than *PS* and its complement) of *PS* will turn up as the probability of an open sentence of *L* and betoken, as I claimed in the Preface, some indefiniteness as to the exact subject matter of that sentence.[20]

A different account of the matter will be found in William Feller, Andrei N. Kolmogorov, Michel Loève, Emanuel Parzen, and others. The writers in question first request, in an informal aside, that a set consist of events of a certain sort, namely, the possible outcomes of such experiments as tossing coins, drawing balls from an urn, and so on, if the set is to qualify as a probability set. They then ask us, given an experiment *E* and the set *PS* of all its possible outcomes, to think of

(1′) the weight of a member, say *O*, of *PS* as the probability (in a presystematic sense of the word) of *E* having outcome *O* as its actual outcome,
(2′) the probability of a subset, say *A*, of *PS* as the probability (in a presystematic sense of the word) of *E* having one of the outcomes in *A* as its actual outcome, and
(3′) the probability of one subset, say *A*, of *PS* given another subset, say *B*, of *PS* as the probability (in a presystematic sense of the word) of *E* having one of the outcomes in *A* as its actual outcome given that *E* has one of the outcomes in *B* as its actual outcome.[21]

Attractive though the account may sound, it has, in my opinion, two serious shortcomings.

First, the account does not suit many of the sets which are acknowledged in the literature as probability sets and hence gives us no clue as to what can be meant in such cases by the probability of a set. It does not suit, for example, the set of all the faces of a given die, a set which Neyman acknowledges in *Lectures and Conferences on Mathematical Statistics and Probability* as a probability set; it does not suit the set of all the thirty-year-old residents of the United States, a set which Ernest Nagel acknowledges in *Principles of the Theory of Probability* as a probability set; and so on.[22] Neyman's account, by contrast, suits all the sets that one can come by, whether they consist of the possible outcomes of a given experiment, the faces of a given die, or the thirty-year-old residents of this country.

Second, the account makes of the figures of Section 6 inductive rather than statistical probabilities. When an experiment *E* is performed, there is nothing indefinite in its having as its actual outcome a given member, say *O*, of the set *PS* of all its possible outcomes; there can be only uncertainty as to whether *E* will as a matter of fact have *O* as its actual outcome or, to switch from events to sentence, uncertainty as to the exact truth-value of the closed sentence '*E* has outcome *O* as its actual outcome.' But uncertainty of this sort is to be measured, I suggested in the Preface, by inductive rather than statistical means—hence my claim that the account makes of the probability of a set an inductive rather than a statistical probability.[23] Neyman's account, by contrast, is unequivocally statistical.

So much then for the rival account. What are we to understand, though, by the probability of an unspecified member of a probability set *PS* belonging to a subset *A* of *PS*? In answer, I shall invite the reader to join in a guessing game of the simplest sort. Suppose we are successively presented with each member of *PS* and asked whether or not it belongs to *A*. Suppose next we choose to answer the question each time in the affirmative. Suppose then we get 0 for each incorrect answer and the weight of the member of *PS* concerned for each correct one. Suppose finally we get as an over-all score for the game: (1) when *PS* is finite, the combined scores we got for our individual answers, (2) when *PS* is denumerably infinite and serially ordered, the limit (if any) for increasing *i* of the combined scores we got for our first *i* answers. The over-all score in question will obviously be equal to the probability of an unspecified member of *PS* belonging to *A*. We may thus understand by this probability the measure of success we would meet with if in the course of the above game we guessed each and every member of *PS*, ir espectively, to belong to *A*. Other interpretations are doubtless possible. The present one, however, will be handy when I study the inferential use to which statistical probabilities may be put.[24]

My first illustration is a time-honored one. Imagine two coins are tossed either simultaneously or in succession, each coin being tossed once and once only. There are obviously four ways in which this simple experiment may terminate: both coins may land heads up (O_1), the first coin may land tails up and the second heads up (O_2), the first coin may land heads up and the second tails up (O_3), and both coins may land tails up (O_4). I shall let my first probability set consist of the four possible outcomes O_1, O_2, O_3, and O_4 of the experiment and show how probabilities may be allotted to its 16 ($=2^4$) subsets by each one of the two procedures sketched on page 34.

According to the first procedure, weights, respectively designated here by '$\mathbf{w}(O_1)$,' '$\mathbf{w}(O_2)$,' '$\mathbf{w}(O_3)$,' and '$\mathbf{w}(O_4)$,' are first allotted to the four members of $\{O_1, O_2, O_3, O_4\}$. Probabilities are then allotted to the various subsets of $\{O_1, O_2, O_3, O_4\}$, the probabilities in question being the combined weights of the members of $\{O_1, O_2, O_3, O_4\}$ which belong to those subsets. Results are as in the following table, where '**Ps**' is short for 'The statistical probability of':

TABLE I

$\mathbf{Ps}(\overline{\{O_1, O_2, O_3, O_4\}}) = 0$	$\mathbf{Ps}(\{O_2, O_3\}) = \mathbf{w}(O_2) + \mathbf{w}(O_3)$
$\mathbf{Ps}(\{O_1\}) = \mathbf{w}(O_1)$	$\mathbf{Ps}(\{O_2, O_4\}) = \mathbf{w}(O_2) + \mathbf{w}(O_4)$
$\mathbf{Ps}(\{O_2\}) = \mathbf{w}(O_2)$	$\mathbf{Ps}(\{O_3, O_4\}) = \mathbf{w}(O_3) + \mathbf{w}(O_4)$
$\mathbf{Ps}(\{O_3\}) = \mathbf{w}(O_3)$	$\mathbf{Ps}(\{O_1, O_2, O_3\}) = \mathbf{w}(O_1) + \mathbf{w}(O_2) + \mathbf{w}(O_3)$
$\mathbf{Ps}(\{O_4\}) = \mathbf{w}(O_4)$	$\mathbf{Ps}(\{O_1, O_2, O_4\}) = \mathbf{w}(O_1) + \mathbf{w}(O_2) + \mathbf{w}(O_4)$
$\mathbf{Ps}(\{O_1, O_2\}) = \mathbf{w}(O_1) + \mathbf{w}(O_2)$	$\mathbf{Ps}(\{O_1, O_3, O_4\}) = \mathbf{w}(O_1) + \mathbf{w}(O_3) + \mathbf{w}(O_4)$
$\mathbf{Ps}(\{O_1, O_3\}) = \mathbf{w}(O_1) + \mathbf{w}(O_3)$	$\mathbf{Ps}(\{O_2, O_3, O_4\}) = \mathbf{w}(O_2) + \mathbf{w}(O_3) + \mathbf{w}(O_4)$
$\mathbf{Ps}(\{O_1, O_4\}) = \mathbf{w}(O_1) + \mathbf{w}(O_4)$	$\mathbf{Ps}(\{O_1, O_2, O_3, O_4\}) = \mathbf{w}(O_1) + \mathbf{w}(O_2) + \mathbf{w}(O_3)$ $+ \mathbf{w}(O_4) = 1$

According to the second procedure, probabilities are directly allotted to the various subsets of $\{O_1, O_2, O_3, O_4\}$. The job proves to be simpler than it may sound. Observe first that, since the two sets $\{O_1, O_2, O_3, O_4\}$ and $\overline{\{O_1, O_2, O_3, O_4\}}$ do not overlap,

$$\mathbf{Ps}(\{O_1, O_2, O_3, O_4\} \cup \overline{\{O_1, O_2, O_3, O_4\}})$$

is equal to

$$\mathbf{Ps}(\{O_1, O_2, O_3, O_4\}) + \mathbf{Ps}(\overline{\{O_1, O_2, O_3, O_4\}}).$$

Since, on the other hand, the two sets $\{O_1, O_2, O_3, O_4\} \cup \overline{\{O_1, O_2, O_3, O_4\}}$ and $\{O_1, O_2, O_3, O_4\}$ are the same,

$$\mathbf{Ps}(\{O_1, O_2, O_3, O_4\} \cup \overline{\{O_1, O_2, O_3, O_4\}})$$

is equal to

$$\mathbf{Ps}(\{O_1, O_2, O_3, O_4\}).$$

Thus $\mathbf{Ps}(\overline{\{O_1, O_2, O_3, O_4\}})$ has to be 0.[25] Observe next that, since the two sets $\{O_1\}$ and $\{O_2\}$ do not overlap, $\mathbf{Ps}(\{O_1, O_2\})$ has by D5.11(b) to be $\mathbf{Ps}(\{O_1\}) + \mathbf{Ps}(\{O_2\})$; that, since the two sets $\{O_1, O_2\}$ and $\{O_3\}$ do not overlap, $\mathbf{Ps}(\{O_1, O_2, O_3\})$ has by D5.11(b) again to be $\mathbf{Ps}(\{O_1, O_2\}) + \mathbf{Ps}(\{O_3\})$; and so on. Observe finally that $\mathbf{Ps}(\{O_1, O_2, O_3, O_4\})$ has to be 1. Once probabilities have been allotted to $\{O_1\}$, $\{O_2\}$, $\{O_3\}$, and $\{O_4\}$, the rest of the job can therefore be handled as in the following table:

TABLE II

$\mathbf{Ps}(\overline{\{O_1, O_2, O_3, O_4\}}) = 0$

$\mathbf{Ps}(\{O_1, O_2\}) = \mathbf{Ps}(\{O_1\}) + \mathbf{Ps}(\{O_2\})$

$\mathbf{Ps}(\{O_1, O_3\}) = \mathbf{Ps}(\{O_1\}) + \mathbf{Ps}(\{O_3\})$

$\mathbf{Ps}(\{O_1, O_4\}) = \mathbf{Ps}(\{O_1\}) + \mathbf{Ps}(\{O_4\})$

$\mathbf{Ps}(\{O_2, O_3\}) = \mathbf{Ps}(\{O_2\}) + \mathbf{Ps}(\{O_3\})$

$\mathbf{Ps}(\{O_2, O_4\}) = \mathbf{Ps}(\{O_2\}) + \mathbf{Ps}(\{O_4\})$

$\mathbf{Ps}(\{O_3, O_4\}) = \mathbf{Ps}(\{O_3\}) + \mathbf{Ps}(\{O_4\})$

$\mathbf{Ps}(\{O_1, O_2, O_3\}) = \mathbf{Ps}(\{O_1\}) + \mathbf{Ps}(\{O_2\}) + \mathbf{Ps}(\{O_3\})$

$\mathbf{Ps}(\{O_1, O_2, O_4\}) = \mathbf{Ps}(\{O_1\}) + \mathbf{Ps}(\{O_2\}) + \mathbf{Ps}(\{O_4\})$

$\mathbf{Ps}(\{O_1, O_3, O_4\}) = \mathbf{Ps}(\{O_1\}) + \mathbf{Ps}(\{O_3\}) + \mathbf{Ps}(\{O_4\})$

$\mathbf{Ps}(\{O_2, O_3, O_4\}) = \mathbf{Ps}(\{O_2\}) + \mathbf{Ps}(\{O_3\}) + \mathbf{Ps}(\{O_4\})$

$\mathbf{Ps}(\{O_1, O_2, O_3, O_4\}) = 1.$

Picturesque readings of the above probabilities will be found in the literature. $\mathbf{Ps}(\{O_1, O_3\})$, for example, the probability of an unspecified one of the possible outcomes O_1, O_2, O_3, and O_4 of the experiment being O_1 or O_3, is often called the probability of the first coin landing heads up; $\mathbf{Ps}(\{O_3, O_4\})$, the probability of an unspecified one of the possible outcomes O_1, O_2, O_3, and O_4 of the experiment being O_3 or O_4, is often called the probability of the second coin landing tails up; and so on.

Probabilities may also be allotted to the various pairs of subsets of $\{O_1, O_2, O_3, O_4\}$ by the procedure of pages 35–36. For example, $\mathbf{Ps}(\{O_1\}, \{O_1, O_3\})$, the so-called probability of the second coin landing heads up given that the first coin has done so, would thereby be equal to

$$\frac{\mathbf{Ps}(\{O_1\} \cap \{O_1, O_3\})}{\mathbf{Ps}(\{O_1, O_3\})},$$

hence to

$$\frac{\mathbf{Ps}(\{O_1\})}{\mathbf{Ps}(\{O_1, O_3\})},$$ [26]

and hence to

$$\frac{\mathbf{Ps}(\{O_1\})}{\mathbf{Ps}(\{O_1\}) + \mathbf{Ps}(\{O_3\})};$$

$\mathbf{Ps}(\{O_3\}, \{O_1, O_3\})$, the so-called probability of the second coin landing tails up given that the first coin has landed heads up, would thereby be equal to $\mathbf{Ps}(\{O_3\})/(\mathbf{Ps}(\{O_1\}) + \mathbf{Ps}(\{O_3\}))$; and so on.

One might wish to allot weights of $\frac{1}{4}$ to each one of the possible outcomes O_1, O_2, O_3, and O_4 of the experiment or, equivalently, probabilities of $\frac{1}{4}$ to each one of the unit sets of those outcomes. Other allotments are not out of the question, however, as we shall see in Section 8.

I assumed above that each coin was tossed once and once only. Imagine instead that each coin is tossed n ($n \geq 2$) times. Since each pair of tosses may terminate in 4 different ways, the n pairs of tosses may now terminate in 4^n different ways. One probability set could consist of these possible outcomes of the n experiments. Another could consist of the n actual outcomes of the n experiments, a possibility too trivial of course to bother with in the previous case. The reader may be trusted, at this point, to study both probability sets on his own. Let me draw his attention, though, to 4 of the 2^n subsets of the second probability set, the set of all the actual outcomes in which both coins land heads up, the set of all those in which the first coin lands tails up and the second heads up, the set of all those in which the first coin lands heads up and the second tails up, and the set of all those in which both coins land tails up. If equal weights were allotted to the n actual outcomes of the n experiments, the probability of each one of the above subsets would prove to be the fraction of the times the coins land as described throughout the n experiments, a fraction of special interest when equal to $\frac{1}{4}$. The pair of coins could then be said to be fair or, more exactly, to have been fair throughout the n experiments.[27]

No pair of coins anyone knows of has ever been tossed an infinite number of times. Suppose, however, to round out the picture, that a pair of coins were tossed as many times, for example, as there are natural numbers and that the $\aleph_0$ experiments in question were performed in a given serial order. One probability set could consist of the $4^{\aleph_0}$ possible outcomes of the experiments, another of their $\aleph_0$ actual outcomes. The first probability set would fall beyond the scope of this book. The $2^{\aleph_0}$ subsets of the second might, however, be allotted probabilities by the procedure of pages 36–37. Banking on the serial order in which the experiments are supposedly performed, we might, for example, take the probability of a given set of actual outcomes to be the limit (if any) for increasing i of the relative frequency with which the actual outcomes of the first i experiments belong to the set in question. The probability of the set of all the actual outcomes in which both coins land heads up, or the first coin lands tails up and the second heads up, or the first coin lands heads up and the second tails up, or both coins land tails up, would thereby prove to be the limit (if any) for increasing i of the fraction of the times the coins land as described throughout the first i experiments, a limit of special interest when equal to $\frac{1}{4}$. The pair of coins could then be said to have been fair throughout the infinitely many experiments.

When on page 41 I tossed each one of my two coins once and once only, I took my probability set to consist of the following four possible outcomes: both coins landing heads up (O_1), the first coin landing tails up and the second heads up (O_2), the first coin landing heads up and the second tails up (O_3), and both coins landing tails up (O_4). I could have taken it instead to consist of the following four possible outcomes: the first coin landing heads up (O_1'), the first coin landing tails up (O_2'), the second coin landing heads up (O_3'), and the second coin landing tails up (O_4'). The members of $\{O_1, O_2, O_3, O_4\}$ would now turn up as pairs of members of $\{O_1', O_2', O_3', O_4'\}$, namely, O_1 as $\langle O_1', O_3'\rangle$, O_2 as $\langle O_2', O_3'\rangle$, O_3 as $\langle O_1', O_4'\rangle$, and O_4 as $\langle O_2', O_4'\rangle$. Instead of allotting weights to the members of $\{O_1, O_2, O_3, O_4\}$, I would now allot weights to the members and pairs of members of $\{O_1', O_2', O_3', O_4'\}$ and take the probability of a set of members or pairs of members of $\{O_1', O_2', O_3', O_4'\}$ to be the combined weights of the members or pairs of members of $\{O_1', O_2', O_3', O_4'\}$ which belong to the said set. An interesting possibility would arise here. If the four pairs $\langle O_1', O_3'\rangle$, $\langle O_2', O_3'\rangle$, $\langle O_1', O_4'\rangle$, and $\langle O_2', O_4'\rangle$ were respectively allotted as their weights the products $\mathbf{w}(O_1')\cdot\mathbf{w}(O_3')$, $\mathbf{w}(O_2')\cdot\mathbf{w}(O_3')$, $\mathbf{w}(O_1')\cdot\mathbf{w}(O_4')$, and $\mathbf{w}(O_2')\cdot\mathbf{w}(O_4')$, the two tosses could be said to be stochastically independent of each other.[28] Similar remarks hold true of all the other experiments which I conducted with the two coins.

All my illustrations so far have featured events, more specifically, the possible or actual outcomes of tossing two coins. My last one will feature people. Consider the set of all the people who turned 65 in

1925, the set of all those who turned 65 in 1926, the set of all those who turned 65 in 1927, and so on. Assuming that the 65-year-olds in each set are allotted equal weights, we might take the probability of a subset of any one of the sets to be the relative frequency with which the 65-year-olds in that set belong to the subset. We might, for example, take the probability of an unspecified 65-year-old in any one set reaching the age of 66 to be the relative frequency with which the 65-year-olds in that set reach the age of 66.

To idealize matters now, consider the set of all the people who would turn 65 in any one of an infinite number of years. Assuming that the people in question are arranged in a given serial order and reckoning things as in the third paragraph of page 37, we might take the probability of an unspecified one of our 65-year-olds reaching the age of 66 to be the limit (if any) for increasing i of the relative frequency with which the first i of them reach that age. The limit in question is what statisticians have in mind when they talk of the probability of a 65-year-old reaching the age of 66. They do not claim, however, to have actually reckoned it and make do, as one would expect, with an estimate of it.

I should like, on closing this section, to discuss a controversial issue, the so-called single-case issue. We know how to reckon, in principle at least, the probability of an unspecified member of a probability set PS belonging to a subset A of PS, for example, the probability of an unspecified Groton senior being admitted to Harvard. How should we reckon, however, the probability of a member of PS belonging to A when the member is explicitly identified by name? Assuming, for example, that Chauncey van Pelt is a Groton senior, how should we reckon the probability of his being admitted to Harvard?

The issue can be met, I believe, in two different ways. Since Chauncey van Pelt is the one and only Groton senior who is Chauncey van Pelt, we might let the probability of his being admitted to Harvard be that of an unspecified one of all the Groton seniors who are Chauncey van Pelt being admitted to Harvard. The probability in question would of course prove to be 1 or 0, 1 if Chauncey is admitted to Harvard, 0 if he isn't. Or else, discounting Chauncey's identity, we might let the probability of his being admitted to Harvard simply be that of an unspecified Groton senior being admitted to Harvard. The probability in question would now presumably be other than 1 or 0.

Generalizing upon this example, we might let the probability of a given member, say a, of PS belonging to a subset A of PS be that of an unspecified member of the intersection $PS \cap \{a\}$ of PS and $\{a\}$ belonging to A. Or else, discounting the identity of the given member of PS, we might let the probability in question simply be that of an unspecified member of PS belonging to A.

I personally favor the first expedient, though a good many writers prefer the second.[29] There is nothing indefinite, so far as I can tell, in Chauncey's being admitted to Harvard. The statistical probability of his being so should therefore be, in my opinion, a flat 1 or 0. To placate the other camp, though, let me hasten to add that: (1) the coefficient of statistical reliability of the inference 'Chauncey van Pelt is a Groton senior, therefore Chauncey van Pelt will be admitted to Harvard,' and (2) the inductive probability of 'Chauncey van Pelt will be admitted to Harvard' given 'Chauncey van Pelt is a Groton senior' will both prove to be other than 1 or 0. The first matter is treated in Section 17, the second in Chapter 4.

8. ABSOLUTE PROBABILITIES, PROBABILITY SETS, AND WEIGHTS

One item from Section 6 is perhaps new to the reader, the notion of an absolute probability. Familiar though he may be with the conditional probability of one set given another as a reference set, he may never have heard of the absolute probability of a subset A of a probability set PS. By definition, however, the latter probability is equal to the conditional probability of A given PS as a reference set. Take, for example, the set of all Protestants and that subset of the said set which consists of all Baptists; the absolute probability of the subset in question is equal by definition to the conditional probability of the set of all Baptists given the set of all Protestants as a reference set. Absolute probabilities, when viewed from this angle, prove to be old friends after all.

Another item from Section 6 may be new to the reader, the notion of a probability set. What, for example, the reader may have known as the conditional probability of one set given another (as a reference set) turns up in Section 6 as the conditional probability of one subset A of a probability set PS given another subset B of PS. His unfamiliarity with probability sets is understandable enough. The conditional probability of one set A given another set B does not usually vary with PS, the probability set which—for the occasion—A and B are presumed to be subsets of, so long as the members of PS are finite in number and are all allotted equal weights.[30] Mention of PS under such circumstances is thus idle and, hence, frequently dropped, whether wittingly or not. The conditional probability in question may, however, vary with PS when PS is not finite; it may also vary with PS, whether or not PS be finite, when the members of PS or of the appropriate subsets of PS are allotted unequal weights.

A vexing problem arises in view of the last two possibilities. Suppose indeed we are confronted with arbitrary pairs of sets and asked to reckon the conditional probability of the first set in any one pair given the second (or, for that matter, the absolute probability of either set in any one pair). Which probability set or sets should we take the sets in question to be subsets of? There is no hard and fast ruling on the matter or, at least, none that I know of. We might be willing to change probability sets from one pair of sets to the other, in which case the union of the two sets in a given pair, for example, would certainly do. Or we might be anxious to retain the same probability set throughout, in which case the union of all the sets in all the pairs would likewise do.

It is the latter strategy which eventually accounts for most infinite probability sets. Some of the sets we deal with in probability theory—those which consist, for example, of the possible outcomes of actually performed experiments[31] or those which consist of the outcomes, whether possible or supposedly actual, of hypothetical experiments[32] —may well be infinite. On the whole, though, the sets in question are finite. They may be quite large, unfortunately, so large indeed that we cannot ascertain their exact size. Shrugging off minutiae, we accordingly fancy them and hence their union to be infinite. To bolster up our fiction, we must of course pack our sets with ad hoc individuals, the way we may have packed the primitive vocabulary of L^{∞} in Chapter 1 with ad hoc individual constants. Since, however, a probability, when tending to a given limit, gets closer and closer by definition to that limit, this bit of deceit on our part may not significantly affect the figure we get when reckoning the probability of a pair of sets, and it provides us with a master probability set wide enough to shelter any pair of sets we may ever have to contend with.[33]

A last item from Section 6 may be new to the reader. In some primers the members of a probability set *PS* when *PS* is finite, or of appropriate subsets of *PS* when *PS* is infinite, are automatically allotted equal weights. The allotment in question, if discussed at all, is usually upheld as the fairest of them all, a view I share up to a point. There are cases, however, where it seems only proper to discriminate against certain members of a probability set *PS* and hence allot them lesser weights. I study two such cases which are typical of many others.

Let us go back once more to my first illustration, the two coins tossed once and once only. The experiment may be said to generate one probability set, the set $\{O_1, O_2, O_3, O_4\}$ of page 41. It may also be said to generate another probability set, the set $\{O_1', O_2', O_3', O_4'\}$ of page 45, where the pairs $\langle O_1', O_3' \rangle$, $\langle O_2', O_3' \rangle$, $\langle O_1', O_4' \rangle$, and $\langle O_2', O_4' \rangle$ do duty for the four members of the first probability set. Now it is clear that, however the members of $\{O_1, O_2, O_3, O_4\}$ be individually weighted, the remaining twelve of the sixteen pairs in which the members of $\{O_1', O_2', O_3', O_4'\}$ can be arranged will have to be weighted 0 if $\langle O_1', O_3' \rangle$, $\langle O_2', O_3' \rangle$, $\langle O_1', O_4' \rangle$, and $\langle O_2', O_4' \rangle$ are to have the same weights as O_1, O_2, O_3, and O_4, respectively. Here therefore is one instance where, to keep the weight and hence probability of certain outcomes of a given experiment from varying, as it otherwise would, with the probability sets generated by the experiment, one might choose to discriminate against some pairs of members of a probability set.

Suppose next that for some reason or other the two coins we are tossing cannot be told apart. The experiment may still be said to be susceptible of four outcomes—the outcomes O_1, O_2, O_3, and O_4—and hence to generate the probability set $\{O_1, O_2, O_3, O_4\}$. Since, however, two of these outcomes, namely, O_2 and O_3, are indistinguishable from each other, the experiment may also be said to be susceptible of three outcomes: both coins landing heads up (O_1''), one coin landing heads up and the other tails up (O_2''), and both coins landing tails up (O_3''), and hence to generate another probability set, the set $\{O_1'', O_2'', O_3''\}$, in which O_2'' does duty for both O_2 and O_3. Now it is clear that, however the members of $\{O_1'', O_2'', O_3''\}$ be individually weighted, $\mathbf{w}(\{O_2\}) + \mathbf{w}(\{O_3\})$ will have to equal $\mathbf{w}(\{O_2''\})$ and hence each one of $\mathbf{w}(\{O_2\})$ and $\mathbf{w}(\{O_3\})$ will normally have to be smaller than either $\mathbf{w}(\{O_1\})$ or $\mathbf{w}(\{O_4\})$ if $\mathbf{Ps}(\{O_1\})$ is to equal $\mathbf{Ps}(\{O_1''\})$, $\mathbf{Ps}(\{O_2, O_3\})$ to equal $\mathbf{Ps}(\{O_2''\})$, and $\mathbf{Ps}(\{O_4\})$ to equal $\mathbf{Ps}(\{O_3''\})$. Here therefore is another instance where, to keep the probability of sets consisting of certain outcomes of a given experiment from varying, as it otherwise would, with the probability sets generated by the experiment, one might choose to discriminate against some members of a probability set.[34]

I leave it to set theorists to spell out the conditions under which two probability sets may be said to be generated by the same experiment, those under which two members of a probability set PS may be said to be indistinguishable in PS, and so on. The task is an urgent one to which they might well address themselves more fully than they have done so far. I shall, however, list the conditions under which, given two individual constants W_1 and W_2 of L, the individuals designated in L by W_1 and W_2 might be said to be indistinguishable in L. First,

$$G(W_1) \equiv G(W_2)$$

should be true in L for every one-place predicate G of L; second,

$$(\forall W_3)(G(W_1, W_3) \equiv G(W_2, W_3))$$

and

$$(\forall W_3)(G(W_3, W_1) \equiv G(W_3, W_2))$$

should be true in L for every two-place predicate G of L; and so on. The conditions in question, easily extended to suit pairs, triples, and so on, of individual constants of L, will be handy in Chapter 3 when the weights allotted here to the members, pairs of members, triples of members, and so on, of a probability set *PS* are turned over to the individual constants, pairs of individual constants, triples of individual constants, and so on, of some language L^N or other.

Further illustrations will be found in the literature.[35] Some are in the same vein as the above. Others, however, are in a different and possibly more controversial vein. I shall study one of them in Section 11.

9. RANDOM FUNCTIONS

Most probability theorists are pledged, on their own admission, to deal with sets. They soon start talking, though, of the probability of the value of some function or other (a random function they call it)[36] being equal to, smaller than, or larger than a given number, and thereby intimate that they can cope with closed sentences as well as with sets. For Feller, Kolmogorov, Loève, and Parzen this manner of speaking may be entirely proper; it is not, however, fitting for those of Neyman's persuasion. I should accordingly like to turn for a moment to random functions and clear up, if I may, this delicate point.

A function, as the reader undoubtedly knows, is a relation which is borne to any one thing, the so-called argument of the function, by one and only one thing, the so-called value of the function for that argument. A numerical function is a function whose values are numbers. And a random function is a numerical function whose arguments have been allotted weights and hence constitute a probability set.

Since the values of a random function are numbers, we can obviously talk of the set of all the arguments of a given random function, say f, for which the value of f is equal, for example, to a given number. Since the arguments of a random function constitute a probability set, we can go one better and talk of the probability of an unspecified argument of f belonging to the set of all the arguments of f for which the value of f is equal to a given number. But the argument in question will clearly belong to the set in question if and only if f takes the number in question as its value for that argument. We can therefore talk of the probability of the value of f being equal for an unspecified one of its arguments to a given number. This, I believe, is what writers of Neyman's persuasion have in mind when they talk of the probability of the value of a so-called random function being equal to a given number. By dropping the clause 'for an unspecified one of its arguments,' they invite, however, the misconception I complained of.

An illustration may be in order. Imagine that s ($s \geq 1$) balls are drawn from an urn filled, in whatever proportions, with red balls and black ones, and that the possible outcomes of the experiment are styled in the usual fashion as so many selections of balls. We can talk here of the set of all the selections in which the number of red balls is equal to a given number, say s_1. We can also talk of the probability of an unspecified selection belonging to the set of all selections in which the number of red balls is equal to s_1. We can finally talk, using the same shortcut as in the last paragraph, of the probability of the number of red balls in an unspecified selection being equal to s_1.

How probabilities are allotted, so to speak, to random functions should be clear from the foregoing. When the arguments of a given random function, say f, are finite in number, the probability of f being equal for an unspecified one of its arguments to a given number may be taken to be the combined weights of the arguments for which f takes that number as its value; when the arguments are denumerably infinite in number and serially ordered, the probability in question may be taken to be the limit (if any) for increasing i of the combined weights of those among the first i arguments for which f takes the number in question as its value. The probability of the value of f being smaller or larger for an unspecified one of its arguments than a given number may be reckoned along similar lines. I have restricted myself here to one-argument random functions; n-argument ($n \geq 2$) ones raise no new problem.

It is possible to associate with each subset A of a given probability set PS a random function, variously called the indicator function or the characteristic function of A, which for any member of PS as its argument has value 1 when the member belongs to A, value 0 when it does not, and hence to revamp all of statistical probability theory as a study of random functions. I must now turn, however, to a more practical question, the question of randomness.

10. RANDOM SAMPLING AND ATTENDANT DISTRIBUTIONS

A population is nothing but a set with a fancy name. The balls with which the urn of Section 9 was filled, for example, constitute a population; so do the cards in a deck of bridge cards. By a selection or sample from a given population we may understand, on the other hand, either

(*a*) a finite subset of that population, or

(*b*) a finite sequence of members of that population.

We may further distinguish under (*b*) between

(*b*1) a sequence in which repetitions are not allowed, and

(*b*2) a sequence in which repetitions are allowed.

As one familiar with Section 5 will undoubtedly guess, a (*b*1)-sample differs from an (*a*)-sample. The order in which various members of a population enter a sample does not matter with an (*a*)-sample; it does with a (*b*1)-sample. A (*b*2)-sample also differs from an (*a*)-sample. How many times a given member of a population enters a sample does not matter with an (*a*)-sample; it does with a (*b*2)-sample.

Imagine, for illustration's sake, that five cards are drawn in succession from a deck of bridge cards. Two possibilities arise: each card, once drawn from the deck, may be replaced in the deck before the next one is drawn, and then again it may not. In one case the drawing would be said to be done with replacement, in the other without replacement. Consider first the case of five cards drawn without replacement. The cards here will necessarily be distinct from one another and may be treated either as an (*a*)-sample or as a (*b*1)-sample. If we are willing to discount the order in which the five cards were drawn, we may treat them as a mere subset of the population and hence as an (*a*)-sample. If, on the other hand, we insist on taking into account the order in which the five cards were drawn, we must treat them as a sequence of members of the population and hence as a (*b*1)-sample. Consider next the case of five cards drawn with replacement. The cards here need not all be distinct from one another.

If no two cards turn out to be the same, we may treat the five of them either as an (*a*)-sample or, if we insist on taking into account the order in which they were drawn, as a (*b*2)-sample which happened to show no repetitions. If two or more cards turn out to be the same, we must treat the five of them as a (*b*2)-sample.[37]

As the reader may recall from combinatorial analysis, $\binom{p}{s}$ distinct (*a*)-samples of size s ($s \geq 0$), $\binom{p}{s}$ $s!$ distinct (*b*1)-samples of the same size, and p^s distinct (*b*2)-samples of the same size can be drawn from a population of size p ($p \geq s$).[38] Six samples of size 2 can be drawn, for example, from the population consisting of the four individuals a, b, c, and d, namely, $\{a, b\}$, $\{a, c\}$, $\{a, d\}$, $\{b, c\}$, $\{b, d\}$, and $\{c, d\}$. Twelve (*b*1)-samples of the same size can be drawn from that population, namely, $\langle a, b\rangle$, $\langle b, a\rangle$, $\langle a, c\rangle$, $\langle c, a\rangle$, $\langle a, d\rangle$, $\langle d, a\rangle$, $\langle b, c\rangle$, $\langle c, b\rangle$, $\langle b, d\rangle$, $\langle d, b\rangle$, $\langle c, d\rangle$, and $\langle d, c\rangle$. Finally, sixteen (*b*2)-samples of the same size can be drawn from that population, namely, $\langle a, a\rangle$, $\langle b, b\rangle$, $\langle c, c\rangle$, $\langle d, d\rangle$, plus the twelve (*b*1)-samples just listed.

Two fundamental theorems are provable at this juncture. Suppose a population of size p is partitioned into two cells of sizes p_1 ($0 \leq p_1 \leq p$) and $p - p_1$, respectively; suppose an (*a*)-sample or a (*b*1)-sample of size s is drawn from the population; and suppose the possible outcomes of the drawing, $\binom{p}{s}$ of them in the first case, $\binom{p}{s}$ $s!$ of them in the second, are allotted equal weights. The probability that in an unspecified one of the outcomes s_1 items hail from the first cell of the population[39] (and hence $s - s_1$ from the second) can then be shown to be

$$\binom{p_1}{s_1}\binom{p - p_1}{s - s_1}\bigg/\binom{p}{s},$$

a ratio known as the hypergeometric distribution and to be referred to here as $H[s_1;p,p_1,s$ (Theorem I)].[40] Suppose, on the other hand, a (*b*2)-sample of size s is drawn from the population in question; suppose also the possible outcomes of the drawing, p^s of them, are allotted equal weights. The probability that in an unspecified one of the outcomes s_1 items hail from the first cell of the population can then be shown to be

$$\binom{s}{s_1}(p_1/p)^{s_1}(1 - p_1/p)^{s-s_1},$$

a ratio known as the binomial distribution and to be referred to here as $B(s_1;p,p_1,s)$ (Theorem II).[41]

To illustrate matters, suppose the four individuals a, b, c, and d fall into two cells of size 2; suppose an (a)-sample of size 2 is drawn from that population; and suppose the six possible outcomes $\{a, b\}$, $\{a, c\}$, $\{a, d\}$, $\{b, c\}$, $\{b, d\}$, and $\{c, d\}$ of the drawing are allotted equal weights. The probability that in an unspecified one of the outcomes s_1 items hail from the first cell of the population is then $\frac{1}{6}$ when $s_1 = 0$, $\frac{2}{3}$ when $s_1 = 1$, and $\frac{1}{6}$ again when $s_1 = 2$.

Like theorems hold with the actual outcomes of n ($n \geq 2$) drawings substituting for the possible outcomes of a single drawing. Suppose a population of size p is partitioned again into two cells of sizes p_1 and $p - p_1$, respectively; suppose n ($n \geq 2$) samples of size s are drawn from the population (each sample being replaced in the population before the next one is drawn) and weighted alike; and suppose the various kinds of sample of size s that can be drawn from the population are evenly represented among the n samples.[42] The probability that in an unspecified one of the n samples s_1 items hail from the first cell of the population can then be shown to be $H(s_1;p,p_1,s)$ when the samples are (a)-samples or ($b1$)-samples, and $B(s_1;p,p_1,s)$ when they are ($b2$)-samples.

To carry on the previous illustration, suppose 180 (a)-samples of size 2 are drawn from the population consisting of a, b, c, and d, and are allotted equal weights; suppose also each one of $\{a, b\}$, $\{a, c\}$, $\{a, d\}$, $\{b, c\}$, $\{b, d\}$, and $\{c, d\}$ turns up 30 times among the 180 samples. The probability that in an unspecified one of the 180 samples s_1 items hail from the first cell of the population is then $\frac{1}{6}$ when $s_1 = 0$, $\frac{2}{3}$ when $s_1 = 1$, and $\frac{1}{6}$ again when $s_1 = 2$.

Like theorems hold, finally, with the actual outcomes of a denumerable infinity of drawings substituting for the possible outcomes of a single drawing. Details are left to the reader.

When it comes to stating Theorems I and II, the literature often uses a locution of its own—the locution 'drawn at random' —which I must now define.[43] Suppose a sample of size s is drawn from a population of size p. The sample in question will be said to be drawn at random from its parent population if the possible outcomes of the drawing—$\binom{p}{s}$ of them when the sample is an (a)-sample, $\binom{p}{s} s!$ of them when the sample is a ($b1$)-sample, and p^s of them when the sample is a ($b2$)-sample—are all weighted alike.[44] With this definition on hand our two theorems can obviously be amended to read: "If a sample of size s is drawn at random from a population of size p partitioned into two cells of sizes p_1 and $p - p_1$, respectively, the probability that in an unspecified one of the possible outcomes of the drawing s_1 items hail from the first cell of the population is equal to $H(s_1;p,p_1,s)$ when the sample is an (a)-sample or a ($b1$)-sample, and to $B(s_1;p,p_1,s)$ when the sample is a ($b2$)-sample" [45] (Theorems I′-II′).

The very same locution could be used to reformulate our other theorems. Suppose indeed n ($n \geq 2$) samples of size s are drawn from a population of size p, each sample being replaced in the population before the next one is drawn. By analogy with the above case the samples could be said to be drawn at random from their parent population if (1) they were all weighted alike and (2) the various kinds of sample of size s that can be drawn from the population were evenly represented among them.[46] Or suppose a denumerable infinity of samples of size s were drawn in succession from a population of size p, each sample being replaced in the population before the next one is drawn. By analogy again with the above case, the samples could be said to be drawn at random from their parent population if (1′) they were drawn in some serial order, (2′) the first i of them were, for each i from 1 on, allotted equal weights, and (3′) the various kinds of sample of size s that can be drawn from the population tended, as more and more samples were drawn, to be evenly represented among them. With these two definitions on hand our other theorems could obviously be amended to read: "If n ($n \geq 2$) samples of size s or a denumerable infinity of samples of size s are drawn at random from a population of size p partitioned into two cells of sizes p_1 and $p - p_1$, respectively, the probability that in an unspecified one of the samples s_1 items hail from the first cell of the population is equal to $H(s_1;p,p_1,s)$ or to $B(s_1;p,p_1,s)$, as the case may be."

Of the three notions of random drawing I have just studied—random drawing of a single sample, random drawing of n ($n \geq 2$) samples, and random drawing of a denumerable infinity of samples—the first one has little, if any, empirical flavor. In much of the literature, however, each one of the possible outcomes of a single drawing is presumed to have as its weight the fraction of the times it has turned up among the actual outcomes of repeated drawings from that population. The presumption lends the first notion some of the empirical flavor of the last two; it does, however, raise a problem which I shall discuss in the next section.

11. STATISTICAL PROBABILITIES ESTIMATED BY MEANS OF SAMPLE SOUNDINGS

Instructions were given in Section 6 for reckoning the probability of two subsets, say A and B, of a finite probability set PS. They read, in brief: add the weights of the members of PS which belong to both A and B, add the weights of the members of PS which belong to B, and divide the first figure by the second; or, when equal weights are allotted to the members of PS, count how many members of PS belong to both A and B, count how many members of PS belong to B, and divide the first figure by the second. Simply worded though the instructions may be, they are unfortunately not always easy to follow.

Let us assume, to simplify matters at the moment, that the members of PS are allotted equal weights. The two subsets of PS we are concerned with may be given by enumeration (as in the first illustration of Section 7), and then again they may not (as in the remaining illustrations of that section). Reckoning the probability of the two subsets in the first case should present no difficulty. Reckoning it in the second, however, may prove to be practically unfeasible either because the second subset is not wholly available for inspection (as when it consists of the past and the future outcomes of an experiment) or because the second subset, though wholly available for inspection, is too large or too costly to inspect. We must then make do with an estimate of the probability in question, that is, with a figure which may well differ from the correct one but nonetheless constitutes an acceptable surrogate for the correct one.

I shall study in Chapter 4 a family of functions whose values for a sentence P of L and a closed sentence Q of L qualify as estimates—made in the light of Q—of the (statistical) probability of P in L and hence, by extension, of the set designated in L by the set abstract $\hat{W}P$ of L. Statisticians, however, have a way of their own of estimating the probability of a pair of sets which runs roughly as follows. Given two subsets A and B of a finite probability set PS or, for that matter, of an infinite probability set PS, they draw from B a sample of some size s (as large an s usually as they can afford); they then count the number, call it s_1, of items in the sample which belong to A; they finally take the ratio s_1/s to be their estimate of the probability of A given B.[47]

The estimate in question is often said to have impressive credentials. The probability that in the sample drawn from B exactly s_1 items hail from A is often claimed to be at its largest (or nearly so) when the relative frequency with which the members of B belong to A equals s_1/s.[48] The claim, however, must be qualified. First, what is at its largest, when the relative frequency with which the members of B belong to A equals s_1/s, is not the probability that in the sample drawn from B exactly s_1 items hail from A;[49] it is rather the probability that in an unspecified one of the possible outcomes of drawing a sample from B or in an unspecified one of the actual outcomes of drawing finitely or infinitely many samples from B exactly s_1 items hail from A. Second, if the probability in question is to be at its largest when the relative frequency with which the members of B belong to A equals s_1/s, the outcomes in question must be allotted the same weights or, to put it more picturesquely, the drawing of the sample or samples in question must be carried out at random. The two qualifications are serious ones and must be placed upon all the claims currently advanced for s_1/s.[50]

If the probability of a pair of sets is to be estimated in the above manner, at least part of the second set must be available for inspection. This, however, often fails to be the case, as when the second set consists of the future outcomes of an experiment. The current practice then, among many statisticians, is to turn to a fresh pair of sets which bears some analogy to the original one and draft its probability (be it the actual or the estimated probability of the pair) as an estimate of the probability of the original pair. The probability of the second one of two coins landing heads up given that the first one has, for example, would often be estimated to be the number of times both coins have landed heads up in the past over the number of times the first one has landed heads up or, if the two coins have never been tossed before, the number of times some two similar coins have landed heads up in the past over the number of times the second one of the said coins has landed heads up. This second manner of estimating the probability of a pair of sets may have credentials of its own, but they have never been examined in much detail in the literature.

I presumed above that the members of *PS* were allotted equal weights. When the members in question are allotted unequal weights, estimating the probability of a pair of sets by either one of the above methods proves to be an extremely delicate affair which statisticians have little studied so far and into which I cannot go here.

Sensible though current methods of estimating the probability of a pair of sets and, as a result, of a set may be, the figures so obtained should not (as, in my opinion, happens all too frequently) be passed off as weights. The point is well worth looking into; it will also permit me to comment anew on the allotting of unequal weights to the members of various probability sets.

Imagine a coin has been tossed 100 times, has landed heads up 40 times, and is about to be tossed 100 more times. Most statisticians would estimate at $\frac{4}{10}$ the fraction of the times the coin will have landed heads up by the close of the whole experiment and hence estimate at 0.4 the probability of the set of those among the actual outcomes of the 200 tosses in which the coin has landed or will land heads up. Many of them would even go one better and set at 0.4 the weight of the coin landing heads up if tossed once more on some future occasion, thus ending up with a probability set with members of unequal weights. Had the coin, on the other hand, landed heads up 50 instead of 40 times in the course of the initial 100 tosses, the weight of its landing heads up when tossed once more on some future occasion would commonly be taken to be 0.5 and, the two possible outcomes of the experiment being allotted equal weights, the coin would be said to be fair.

The drawing of samples fares in the literature like the tossing of coins. Imagine indeed that 180 (*a*)-samples of size 2 have been drawn from the population consisting of the four individuals *a*, *b*, *c*, and *d*; imagine next that out of the six samples of size 2 that can be drawn from the said population five have turned up 25 times each in the course of the 180 drawings and the sixth one 55 times; imagine finally that a new sample of size 2 is to be drawn from the population. Many statisticians would set at $\frac{5}{36}$ the weight of five of the possible outcomes of the experiment and at $\frac{11}{36}$ the weight of the sixth one, thus ending up again with a probability set with members of unequal weights. Had the six samples, on the other hand, turned up 30 times each in the course of the 180 drawings, the six possible outcomes of the experiment would commonly be allotted equal weights and the drawing of the new sample would commonly be said to be carried out at random.

This legerdemain, whereby an estimate of a probability, the probability of a set of actual outcomes, is made to serve as a weight, the weight of a possible outcome, seems to me open to question.[51] Statistical probabilities and hence weights should not, in my opinion, be advertised on one page as measurements on sets and prove on the next to be predictions of such measurements. True, the measurements on sets we call statistical probabilities may frequently be inaccessible to us and hence have to be surmised in some fashion or other. But surmises of one measurement, say the statistical probability of a set of actual outcomes, should not be passed off, at least by those of Neyman's persuasion, as another measurement, say the weight of a possible outcome. If my objection be well taken, the only remaining grounds for allotting unequal weights to the members of a probability set or, at least, the only ones I know of from the literature, would be those I studied in Section 8.

What is to become, though, of the weights allotted a few paragraphs back as weights to the possible outcomes of such experiments as tossing a coin, drawing a sample from a population, and so on? Two suggestions come to mind: (1) When they set at 0.4, for example, the weight of a coin landing heads up, statisticians could be said to be reckoning the probability of an experiment having as its actual outcome a given one of its possible outcomes, and hence to have merely pinned onto 0.4 the wrong one of the two labels 'statistical probability' and 'inductive probability.' (2) Or they could be said to be dealing with a so-called single case, to treat the probability of the coin landing heads up on a given toss as the probability of its landing heads up on an unspecified one of certain tosses (presumably the past and future tosses of the coin), and, being in the dark as to what the latter probability is, to have drafted another figure (the fraction of the times the coin has landed heads up on recent tosses) as a surrogate. The first suggestion is simpler; the second, however, has the merit of squaring 0.4 with Neyman's understanding of weights and probabilities.

NOTES

[1] Probability sets are also known as probability spaces, sample spaces, and so on.

[2] The numbers in question are frequently called probabilities (also elementary probabilities), but a fresh label like 'weight' makes, I believe, for greater clarity.

[3] As announced earlier, I let the four letters '*A*,' '*B*,' '*C*,' and '*D*' range in this chapter over sets.

[4] The first procedure is used by W. Feller, *An Introduction to Probability Theory and Its Applications*, p. 22; it also appears in A. N. Kolmogorov, *Foundations of the Theory of Probability*, p. 3, and E. Parzen, *Modern Probability Theory and Its Applications*, p. 24. Both probabilities and weights are often known, by the way, as measures, and probability theory often treated as a chapter of measure theory.

[5] The second procedure is used by Kolmogorov, *loc. cit.*, p. 2, M. Loève, *Probability Theory*, p. 8, and Parzen, *loc. cit.*, p. 18. Sets which are allotted probabilities are normally required to constitute a field, that is, a set of sets to which

belong, first, the complement of any member set and, second, the intersection (or, equivalently, union) of any two member sets. I informally assume that the subsets of any probability set considered here meet this requirement.

[6] The label 'equiprobable' is used by J. G. Kemeny, H. Mirkil, J. L. Snell, and G. L. Thompson, *Finite Mathematical Structures*, pp. 120–122, the label 'finite-frequency' used by B. Russell, *Human Knowledge, Its Scope and Limits*, pp. 350–362.

[7] For proof that $2^{\aleph_0}$ multiplied by 1 or by itself $n - 1$ times is equal to $2^{\aleph_0}$, see Fraenkel, *loc. cit.*, §§6–7.

[8] For further information on this matter, see Kolmogorov, *loc. cit.*, Chapter II, and Loève, *loc. cit.*, pp. 15–17.

[9] The probability in question may of course vary with the serial order in which the members of *PS* are arranged; see Section 5.

[10] See, for example, E. Nagel, *Principles of the Theory of Probability*, pp. 21–22. Nagel, like many writers, has a so-called property do duty for the set *A*. For another allotment of probabilities which has also been called a relative frequency allotment, see Note 51 of this chapter.

[11] See von Mises, "Grundlagen der Wahrscheinlichkeitsrechnung" (or the more popular *Probability, Statistics, and Truth*), and Reichenbach, *The Theory of Probability*.

[12] See, for example, Nagel, *loc. cit.*, pp. 19–37, and Russell, *loc. cit.*, pp. 362–372

[13] For a brief study of this point, see Nagel, *loc. cit.*, pp. 31–33.

[14] For further details on the matter, see M. E. Munroe, *Theory of Probability*, Chapter 2, and Kemeny, Mirkil, Snell, and Thompson, *loc. cit.*, Chapter 7.

[15] When a pair of members of *PS*, or a triple of members of *PS*, and so on, is allotted as its weight the product of the weights of the two members of *PS*, or the three members of *PS*, and so on, of which it is made up, the members of *PS* in question are usually said to be stochastically independent of one another. See on this point S. Goldberg, *Probability, An Introduction*, pp. 107–121.

[16] A set of pairs of members of *PS* is often said to be a subset of the so-called Cartesian product $PS \times PS$, a set of triples of members of *PS* to be a subset of the so-called Cartesian product $PS \times PS \times PS$, and so on, $PS \times PS$ being the set of all pairs of members of *PS*, $PS \times PS \times PS$ the set of all triples of members of *PS*, and so on. Allotting probabilities to sets of pairs, or triples, and so on, of members of *PS* is therefore simply a matter of allotting probabilities to the subsets of $PS \times PS$, or $PS \times PS \times PS$, and so on, as probability sets.

[17] See *Lectures and Conferences on Mathematical Statistics and Probability*, pp. 2–5 and 13–15. Neyman, who refers to the members of *PS* as objects *A*, talks of the probability of an object *A* having a property *B*; the subsets of *PS*, however, play the same part here as properties do in Neyman.

[18] As Russell does, *loc. cit.*, p. 352.

[19] I shall likewise think of the weight of a pair of members, say $\langle a, b \rangle$, of *PS* as the probability of an unspecified pair of members of *PS* being $\langle a, b \rangle$, of the weight of a triple of members, say $\langle a, b, c \rangle$, of *PS* as the probability of an unspecified triple of members of *PS* being $\langle a, b, c \rangle$, and so on.

[20] To slip in an example, let $\hat{w}$(w outlives Peter) be a subset of some probability set *PS* or other, and let the defining condition 'w outlives Peter' of $\hat{w}$(w outlives Peter) be a sentence of some language *L* or other. The probability of $\hat{w}$(w outlives Peter) will turn up in Chapter 3 as the probability of 'w outlives Peter', an open sentence in which 'w', referring as it does to an unspecified member of $\vee^1_L$, will do duty for Neyman's indefinite article '*a*'. For further remarks on the part played in Neyman by the said article, see H. E. Kyburg, Jr., *Probability and the Logic of Rational Belief*, Chapter 2.

[21] See Feller, *loc. cit.*, pp. 7–9, Kolmogorov, *loc. cit.*, pp. 3–5, Loève, *loc. cit.*, pp. 5–6, and especially Parzen, *loc. cit.*, pp. 11–12 and 17–18, where the point is most clearly made.

[22] For the first illustration, see p. 5 of Neyman's *Lectures and Conferences*; for the second, see pp. 19–22 of Nagel's *Principles*.

[23] Since the closed sentences '*E* has *O* as its actual outcome,' '*E* has one of the outcomes in *A* as its actual outcome,' and so on, can do duty here for such events as *E* having *O* as its actual outcome, *E* having one of the outcomes in *A* as its actual outcome, and so on, it might further be claimed that the probability functions which Feller, Kolmogorov, Loève, Parzen, and others study differ from the inductive probability functions of Chapter 4 only in the restricted kind of sentences they can take as their arguments. The claim (inspired, as it happens, by comments of Professor Carnap and Professor Hermes) will receive some corroboration in Section 11, where (1) I show that the writers in question usually set the weight of a given outcome *O* at the fraction of the times *O* has proved to be, in the course of repeated performances of the experiment *E*, the actual outcome of *E*, and (2) I argue that the fraction in question is best viewed as an estimate of a statistical probability and hence as an inductive probability.

[24] The interpretation offered in the text grew out of a conversation with R. Trent Sorenson, a former graduate student of mine at Bryn Mawr College.

[25] It follows by the very same argument that the complement of any probability set (and hence $\wedge$ when the set $\vee$ of Chapter 1 is appointed to serve as a probability set) has 0 as its probability.

[26] Note for proof that $\{O_1\}$ is the only member of $\{O_1, O_2, O_3, O_4\}$ which belongs to each one of $\{O_1\}$ and $\{O_1, O_3\}$; by D5.2 and D5.6 $\{O_1\} \cap \{O_1, O_3\}$ is therefore identical with $\{O_1\}$.

[27] In much of the literature (1) a coin is said to be fair (without qualification) if the two ways in which it may land are weighted alike, and (2) the two ways in which the coin may land are weighted alike if in the course of repeated tosses the

coin has landed as many times heads up as tails up. See Section 11 on this whole matter.

[28] See Note 15 of this chapter. $\langle O_3', O_1' \rangle$ could also substitute here for O_1, $\langle O_3', O_2' \rangle$ for O_2, $\langle O_4', O_1' \rangle$ for O_3, and $\langle O_4', O_2' \rangle$ for O_4.

[29] Some writers do not even notice the difference, stressed by Neyman and others, between Chauncey van Pelt being admitted to Harvard and an unspecified Groton senior being admitted to Harvard; they thus favor the second expedient rather unwittingly.

[30] The probability in question varies from one probability set, say *PS*, to another, say *PS'*, when some member of *PS* that qualifies for membership in A or in B does not belong to *PS'* or vice versa.

[31] Tossing a coin once is a case in point when the infinitely many ways in which the coin may land on edge are taken into account.

[32] Tossing a coin infinitely many times is a case in point.

[33] The reference to L^∞, a few lines back in the text, is not casual. The problem which arises here with pairs of sets also arises with sentences (be it singly or by pairs), and it is solved in this book through the construction of the master language L^∞, a language wide enough to shelter a good many (though not all) of the sentences to which we may want to allot statistical and inductive probabilities.

[34] For a somewhat different, though related, analysis of the matter, see my paper "Statistical and Inductive Probabilities."

[35] See, for example, Feller, *loc. cit.*, pp. 19–22, Kemeny, Mirkil, Snell, and Thompson, *loc. cit.*, pp. 120–122.

[36] Most writers favor 'random variable' over 'random function,' a rather unfortunate predilection. For a more detailed study of random variables, see K. Menger, "Random Variables from the Point of View of a General Theory of Variables," a reference I owe to Professor Nagel.

[37] Only (*a*)-samples can come up for mention in the languages *L*.

[38] For proof consider first a (*b*2)-sample. There are p candidates in the population for first place in the sample, p candidates for second place, $\cdots$, and p candidates for s-th place. p^s distinct (*b*2)-samples of size s can therefore be drawn from a population of size p. Consider next a (*b*1)-sample. There are p candidates in the population for first place in the sample; first place once filled, $p - 1$ candidates left for second place; $\cdots$; and, $(s - 1)$-th place once filled, $p - s + 1$ candidates left for last place. $p\cdot(p - 1)\cdot \cdots \cdot(p - s + 1)$ distinct (*b*1)-samples of size s can therefore be drawn from a population of size p. But $p\cdot(p - 1)\cdot \cdots \cdot (p - s + 1)$ is equal to $p\cdot(p - 1)\cdot \cdots \cdot(p - s + 1)\cdot(p - s)!\cdot s!/s!\cdot(p - s)!$, which is equal to $\binom{p}{s}s!$. $\binom{p}{s}s!$ distinct (*b*1)-samples of size s can therefore be drawn from a population of size p. Consider finally an (*a*)-sample. The s items

of which a given (b1)-sample is made up can enter the sample in $s!$ distinct ways. There are indeed s candidates among those items for first place in the sample; first place once filled, $s - 1$ candidates left for second place; $\cdots$; and, $(s - 1)$-th place once filled, 1 candidate left for s-th place. But $\binom{p}{s} s!$ distinct (b1)-samples of size s, I just proved, can be drawn from a population of size p. $\binom{p}{s} s!/s!$ or $\binom{p}{s}$ distinct (a)-samples of size s can therefore be drawn from a population of size p. The first result can of course be made to read that n things can be arranged in n^2 pairs, n^3 triples, and so on, and the third to read that a set of size n has $\binom{n}{i}$ subsets of size i $(0 \leq i \leq n)$.

[39] Or, to phrase matters as in Section 9, the probability that in an unspecified one of the outcomes the number of items hailing from the first cell of the population is equal to s_1.

[40] s_1 is presumed throughout to lie in the interval $[\max(0, s - p + p_1), \min(s, p_1)]$.

[41] The binomial distribution is also pertinent when it comes to drawing a sample from an infinite population, a topic for which I have no space here. Proofs of the two theorems are left as an exercise to the reader.

[42] For the condition to be met, n must of course be a multiple of $\binom{p}{s}$, $\binom{p}{s} s!$, or p^s, as the case may be.

[43] The notion of random drawing, though related in many ways to von Mises' notion of random occurrence in a set, should not be confused with it. The former presupposes that probabilities have already been allotted to sets, whereas the latter has to do with the allotting of probabilities to sets.

[44] See, for example, Feller, *loc. cit.*, p. 29.

[45] In writers like Feller, Kolmogorov, Loève, and Parzen, Theorems I and II are further amended to read: "If a sample of size s is drawn at random from a population of size p partitioned into two cells of sizes p_1 and $p - p_1$, respectively, then the probability that s_1 of the items in the sample hail from the first cell of the population is equal to $H(s_1; p, p_1, s)$ when the sample is an (a)-sample or a (b1)-sample, and to $B(s_1; p, p_1, s)$ when the sample is a (b2)-sample." Theorems of this sort belong, in my opinion, to inductive probability theory and are provable indeed for all of the inductive probability functions I consider in Chapter 4. See Section 24 for further details on the matter.

[46] In the course of correspondence we exchanged on the matter, Professor Carnap objected to my using the locution 'drawn at random' in connection with these samples. He claimed that if 52 samples of one card each were drawn from a deck of bridge cards and each card in the deck turned up once in the course

of the 52 drawings, one would hardly dub the drawings random ones. I can only answer that, under the circumstances which Professor Carnap describes, a good many writers would allot equal weights to the possible outcomes of the drawing of a 53rd card from the deck and that they could hardly call this a random drawing without so dubbing the previous 52.

[47] s_1/s is known as a *point* estimate of the probability of A given B. *Interval* estimates of that probability are also in vogue.

[48] And, for that reason, s_1/s is called a maximum likelihood estimate of the probability of A given B.

[49] The probability in question could indeed be said to be a flat 1 or 0, 1 if s_1 of the items in the sample hailed from A, and 0 if any other number of items hailed from A; see pages 46–47.

[50] Writers like Feller, Kolmogorov, Loève, and Parzen may of course ignore the first qualification; but they cannot ignore the second.

[51] Parzen, who, as a rule, takes the weight of a given member, say O, of the set PS of all the possible outcomes of an experiment E to be the fraction of the times E has in the past had O as its actual outcome, refers to his allotment of weights as a relative frequency allotment. See Note 12 of this chapter.

3 STATISTICAL PROBABILITIES: PART TWO

In this third chapter I deal with statistical probabilities as measurements on sentences. First, I show how probabilities normally meant for sets may be passed on to five sample sentences of L (Section 12). Then, I give wholesale instructions for allotting statistical probabilities to the sentences of L (Section 13); establish that the probabilities in question are truth-values of a sort (Section 14); and study in some detail one of the allotments endorsed in Section 13, the relative frequency allotment (Section 15). Finally, after reviewing some rival ways of reckoning the statistical probability of a closed sentence of L (Section 16), I consider the inferential use to which statistical probabilities can be put (Section 17).

12. SETS, SENTENCES, AND STATISTICAL PROBABILITIES

Consider first a closed sentence of L, 'Outlives(Mary,Peter),' for example, or, to keep the number of parentheses down to a minimum here, 'Mary outlives Peter.' It follows from D5.1(a), D5.9–10, and D3.2–3 that the set abstract '$\hat{w}$(Mary outlives Peter)' designates either $\vee$ or $\wedge$ in L, $\vee$ when 'Mary outlives Peter' is true in L, $\wedge$ when 'Mary outlives Peter' is false in L. We may accordingly pass on to 'Mary outlives Peter' the probability which would be allotted in Chapter 2 to $\vee$ or $\wedge$, namely, 1 or 0, and take **Ps**(Mary outlives Peter), the statistical probability in L of 'Mary outlives Peter,' to be 1 when 'Mary outlives Peter' is true in L, 0 when 'Mary outlives Peter' is false in L.[1]

Consider next an open sentence of L^N in which one and only one individual variable of L^N is free, 'w outlives Peter,' for example. It follows from D5.1(a) and D3.5 that a member of the universe of discourse of L^N, say Mary, belongs to the set designated in L^N by '$\hat{w}(w$ outlives Peter)' if and only if Mary satisfies 'w outlives Peter' when assigned to the individual variable 'w.' Heeding a proposal of John G. Kemeny's, we may accordingly take $\mathbf{Ps}^N(w$ outlives Peter), the statistical probability in L^N of 'w outlives Peter,' to be the combined weights of all the members of the universe of discourse of L^N which satisfy 'w outlives Peter' when assigned to 'w.'[2] The probability in question answers the one which would be allotted in Chapter 2 to the subset $\hat{w}(w$ outlives Peter) of an N-membered probability set PS.

It also follows from D5.1(a) and D3.2 that the individual designated in L^N by an individual constant of L^N, say 'Mary,' belongs to the set designated in L^N by '$\hat{w}(w$ outlives Peter)' if and only if 'Mary outlives Peter' is true in L^N. As an alternative to Kemeny's proposal, we may accordingly pass on to the N individual constants of L^N the weights normally allotted to the individuals designated by those constants and take $\mathbf{Ps}^N(w$ outlives Peter) to be the sum

$$\sum_{i=1}^{N} (\mathbf{w}^N(W_i) \cdot \mathbf{Ps}^N(W_i \text{ outlives Peter})),$$

where, for each i from 1 to N, W_i is the i-th individual constant of L^N, $\mathbf{w}^N(W_i)$ is the weight of W_i in L^N, and $\mathbf{Ps}^N(W_i$ outlives Peter) is the statistical probability in L^N (1 or 0 by the above reckoning) of the closed sentence

$$W_i \text{ outlives Peter}$$

of L^N. The two proposals are equivalent so far as our languages go. The second, however, is more convenient and will be favored here.[3]

Consider next an open sentence of L^N in which two and only two individual variables of L^N are free, 'w outlives x,' for example. As we just passed on to the individual constants of L^N the weights normally allotted to the individuals designated by those constants, so we may pass on to the various pairs (N^2 in all) made up of any two individual constants of L^N the weights normally allotted to the pairs of individuals designated by those constants and take $\mathbf{Ps}^N(w$ outlives $x)$ to be the sum

$$\sum_{i=1}^{N^2} (\mathbf{w}^N(W_{i_1}W_{i_2}) \cdot \mathbf{Ps}^N(W_{i_1} \text{ outlives } W_{i_2})),$$

where, for each i from 1 to N^2, $W_{i_1}W_{i_2}$ is (according to some arbitrary ordering)[4] the i-th one of the pairs just mentioned, $\mathbf{w}^N(W_{i_1}W_{i_2})$ is the weight of $W_{i_1}W_{i_2}$ in L^N, and $\mathbf{Ps}^N(W_{i_1}$ outlives $W_{i_2})$ is the statistical probability in L^N (1 or 0 again) of the closed sentence

$$W_{i_1} \text{ outlives } W_{i_2}$$

of L^N.[5] The procedure is easily extended to suit open sentences of L^N in which three or more individual variables of L^N are free.

Consider next an open sentence of L^∞—'w outlives Peter' will do again—in which one and only one individual variable of L^∞ is free. For each N from 1 on, the N results of substituting an individual constant of L^N (and hence of L^∞) for 'w' in 'w outlives Peter' all have probabilities of their own in L^∞. We may accordingly take $\mathbf{Ps}^\infty$ (w outlives Peter), the statistical probability in L^∞ of 'w outlives Peter,' to be the limit (if any) for increasing N of the sum

$$\sum_{i=1}^{N} (\mathbf{w}^N(W_i) \cdot \mathbf{Ps}^\infty (W_i \text{ outlives Peter})),$$

where, for each N from 1 on and each i from 1 to N, W_i is the i-th individual constant of L^N, $\mathbf{w}^N(W_i)$ is the weight of W_i in L^N, and $\mathbf{Ps}^\infty$ (W_i outlives Peter) is the statistical probability in L^∞ (1 or 0 again) of the closed sentence

$$W_i \text{ outlives Peter}$$

of L^∞.[6] The probability in question answers the one which would be allotted in Chapter 2 to the subset $\hat{w}$(w outlives Peter) of a denumerably infinite and serially ordered probability set PS.

Consider next an open sentence of L^∞—'w outlives x' will do again—in which two and only two individual variables of L^∞ are free. For each N from 1 on, the N^2 results of substituting an individual constant of L^N (and hence of L^∞) for each one of 'w' and 'x' in 'w outlives x' all have probabilities of their own in L^∞. We may accordingly take $\mathbf{Ps}^\infty$ (w outlives x) to be the limit (if any) for increasing N of the sum

$$\sum_{i=1}^{N^2} (\mathbf{w}^N(W_{i_1}W_{i_2}) \cdot \mathbf{Ps}^\infty \, (W_{i_1} \text{ outlives } W_{i_2})),$$

where, for each N from 1 on and each i from 1 to N, $W_{i_1}W_{i_2}$ is (according again to some arbitrary ordering) the i-th pair made up of any two individual constants of L^N, $\mathbf{w}^N(W_{i_1}W_{i_2})$ is the weight of $W_{i_1}W_{i_2}$ in L^N, and $\mathbf{Ps}^\infty$ (W_{i_1} outlives W_{i_2}) is the statistical probability in L^∞ (1 or 0 again) of the closed sentence

$$W_{i_1} \text{ outlives } W_{i_2}$$

of L^∞.[7] The procedure is easily extended to suit open sentences of L^∞ in which three or more individual variables of L^∞ are free.

The statistical probabilities I have dealt with so far were absolute ones. Consider finally a pair of sentences of L, 'w prefers filter cigarettes' and 'w is a cigarette smoker,' for example. We may, improving upon the procedure of Chapter 2, take the conditional statistical probability in L of 'w prefers filter cigarettes' given 'w is a cigarette smoker' or, for short, $\mathbf{Ps}$(w prefers filter cigarettes, w is a cigarette smoker) to be as follows: (1) the ratio of $\mathbf{Ps}$(w prefers filter cigarettes & w is a cigarette smoker) to $\mathbf{Ps}$(w is a cigarette smoker) when $\mathbf{Ps}$(w is a cigarette smoker) is non-zero, and, (2) 1 when $\mathbf{Ps}$(w is a cigarette smoker) is zero.[8]

13. STATISTICAL PROBABILITIES ALLOTTED TO THE SENTENCES OF THE LANGUAGES L

The above suggestions for allotting statistical probabilities to the sentences of L may be officially put as follows:

D13.1. *For each N and each n from 1 on, clauses of the following kind:*

(1) $W_{1_1}W_{1_2} \cdots W_{1_n}$ has $\mathbf{w}^N(W_{1_1}W_{1_2} \cdots W_{1_n})$ as its weight in L^N,
(2) $W_{2_1}W_{2_2} \cdots W_{2_n}$ has $\mathbf{w}^N(W_{2_1}W_{2_2} \cdots W_{2_n})$ as its weight in L^N,

.

.

$(N^n) W_{N^n_1}W_{N^n_2} \cdots W_{N^n_n}$ has $\mathbf{w}^N(W_{N^n_1}W_{N^n_2} \cdots W_{N^n_n})$ as its weight in L^N,

are presumed to be on hand, where

(a) *$W_{1_1}W_{1_2} \cdots W_{1_n}$, $W_{2_1}W_{2_2} \cdots W_{2_n}$, $\cdots$, and $W_{N^n_1}W_{N^n_2} \cdots W_{N^n_n}$ are in some arbitrary order the various sequences made up of any n individual constants of the sublanguage L^N of L^∞*

and

(b) *$\mathbf{w}^N(W_{1_1}W_{1_2} \cdots W_{1_n})$, $\mathbf{w}^N(W_{2_1}W_{2_2} \cdots W_{2_n})$, $\cdots$, and $\mathbf{w}^N(W_{N^n_1}W_{N^n_2} \cdots W_{N^n_n})$*

are real numbers such that:

(b1) *$0 \leq \mathbf{w}^N(W_{i_1}W_{i_2} \cdots W_{i_n})$ for each i from 1 to N^n*

and

(b2) $\sum_{i=1}^{N^n} \mathbf{w}^N(W_{i_1}W_{i_2} \cdots W_{i_n}) = 1.$

D13.2. (a) *Let P be a closed sentence of L^N. Then $\mathbf{Ps}^N(P)$ equals 1 if P is true in L^N, 0 if P is not true in L^N.*

(b) *Let P be a closed sentence of L^∞. Then $\mathbf{Ps}^\infty(P)$ equals 1 if P is true in L^∞, 0 if P is not true in L^∞.*

(c) *Let P be an open sentence of L^N; let $X_1, X_2, \cdots$, and X_n be the n ($n \geq 1$) individual variables of L^N which are free in P; and for each i from 1 to N^n let $W_{i_1}W_{i_2} \cdots W_{i_n}$ be the i-th of the sequences made up of any n individual constants of L^N and P_i be like P except for containing occurrences of $W_{i_1}, W_{i_2}, \cdots$, and W_{i_n} at all the places where P contains occurrences of $X_1, X_2, \cdots$, and X_n, respectively. Then*

$$\mathbf{Ps}^N(P) = \sum_{i=1}^{N^n} (\mathbf{w}^N(W_{i_1}W_{i_2} \cdots W_{i_n}) \cdot \mathbf{Ps}^N(P_i)).$$

(d) *Let P be an open sentence of L^∞; let X_1, X_2, $\cdots$, and X_n be the n ($n \geq 1$) individual variables of L^∞ which are free in P; and for each N from 1 on and each i from 1 to N^n let $W_{i_1}W_{i_2} \cdots W_{i_n}$ and P_i be as in (c). Then*

$$\mathbf{Ps}^\infty(P) = \mathit{Limit}_{N\to\infty} \sum_{i=1}^{N^n} (\mathbf{w}^N(W_{i_1}W_{i_2} \cdots W_{i_n}) \cdot \mathbf{Ps}^\infty(P_i))$$

if such a limit exists; otherwise $\mathbf{Ps}^\infty(P)$ *has no value.*

D13.3. *Let both* $\mathbf{Ps}(P \,\&\, Q)$ *and* $\mathbf{Ps}(Q)$ *have a value.*

(a) *If* $\mathbf{Ps}(Q) \neq 0$, *then* $\mathbf{Ps}(P, Q) = \mathbf{Ps}(P \,\&\, Q)/\mathbf{Ps}(Q)$;

(b) *If* $\mathbf{Ps}(Q) = 0$, *then* $\mathbf{Ps}(P, Q) = 1$.

The following theorems are immediate consequences of D13.1–2. The first, T13.4, converts weights into probabilities. The second, T13.5, helps to reckon probabilities when, for each N from 1 on, each sequence of n ($n \geq 2$) individual constants of L^N is allotted as its weight in L^N the product of the weights in L^N of the n constants in question. The third, T13.6, helps to reckon probabilities when, for each N from 1 on, any two sequences of n ($n \geq 1$) individual constants of L^N are allotted the same weight in L^N and hence each sequence of n individual constants of L^N is allotted $1/N^n$ as its weight in L^N. The allotment considered in T13.6 is, of course, the relative frequency one.

T13.4. *Let $W_1W_2 \cdots W_n$ be a sequence of n ($n \geq 1$) individual constants of L^N and let X_1, X_2, $\cdots$, and X_n be n individual variables of L^N distinct from one another. Then*

$$\mathbf{w}^N(W_1W_2 \cdots W_n) = \mathbf{Ps}^N((\cdots(X_1 = W_1 \,\&\, X_2 = W_2) \,\&\, \cdots) \,\&\, X_n = W_n).$$

Proof: By D3.2, D3.1(c), D3.1(a), and D2.7, $(\cdots(Y_1 = W_1 \,\&\, Y_2 = W_2) \,\&\, \cdots) \,\&\, Y_n = W_n$, where Y_1, Y_2, $\cdots$, and Y_n are n individual constants of L^N, is true in L^N if and only if Y_1 is identical with W_1, Y_2 with W_2, $\cdots$, and Y_n with W_n. Hence by D13.2(a) and D2.4 $\mathbf{Ps}^N((\cdots(Y_1 = W_1 \,\&\, Y_2 = W_2) \,\&\, \cdots) \,\&\, Y_n = W_n)$ equals 1 if Y_1 is identical with W_1, Y_2 with W_2, $\cdots$, and Y_n with W_n; otherwise it equals 0. Hence T13.4 by D13.2(c) and the hypothesis on X_1, X_2, $\cdots$, and X_n.

According to T13.4, the weight in L^N of an individual constant of L^N, say the constant 'Mary,' is equal to the probability in L^N of 'w = Mary' or, if you prefer, to the probability of a member of $\vee^1_{L^N}$ being Mary; the weight in L^N of a pair of individual constants of L^N, say the constants 'Mary' and 'Peter,' is equal to the probability in L^N of 'w = Mary & x = Peter' or, if you prefer, to the probability of a member of $\vee^1_{L^N}$ being Mary and a member of $\vee^1_{L^N}$ being Peter; and so on.

T13.5. *For each N from 1 on let*

(1) $$\mathbf{w}^N(W_1W_2) = \mathbf{w}^N(W_1)\cdot\mathbf{w}^N(W_2),$$

where W_1W_2 is any pair of individual constants of L^N,

(2) $$\mathbf{w}^N(W_1W_2W_3) = \mathbf{w}^N(W_1)\cdot\mathbf{w}^N(W_2)\cdot\mathbf{w}^N(W_3),$$

where $W_1W_2W_3$ is any triple of individual constants of L^N, and so on.

(a) *Let P be a closed sentence of L^N. Then* $\mathbf{Ps}^N(P) = 1$ *if P is true in L^N,* $\mathbf{Ps}^N(P) = 0$ *if P is not true in L^N.*

(b) *Let P be a closed sentence of L^∞. Then* $\mathbf{Ps}^\infty(P) = 1$ *if P is true in L^∞,* $\mathbf{Ps}^\infty(P) = 0$ *if P is not true in L^∞.*

(c1) *Let P be an open sentence of L^N; let X_1 be the one individual variable of L^N which is free in P; and for each i from 1 to N let W_i be the i-th individual constant of L^N and P_i be like P except for containing occurrences of W_i at all the places where P contains free occurrences of X_1. Then*

$$\mathbf{Ps}^N(P) = \sum_{i=1}^{N} (\mathbf{w}^N(W_i)\cdot\mathbf{Ps}^N(P_i)).$$

(c2) *Let P be an open sentence of L^N; let X_1, X_2, $\cdots$, and X_{n+1} be the $n + 1$ ($n \geq 1$) individual variables of L^N which are free in P; and for each i from 1 to N let W_i be the i-th individual constant of L^N and P_i be like P except for containing occurrences of W_i at all the places where P contains free occurrences of X_{n+1}. Then*

$$\mathbf{Ps}^N(P) = \sum_{i=1}^{N} (\mathbf{w}^N(W_i)\cdot\mathbf{Ps}^N(P_i)).$$

(d1) *Let P be an open sentence of L^∞; let X_1 be the one individual* *variable of L^∞ which is free in P; and for each i from 1 on let W_i be the i-th individual constant of L^∞ and P_i be like P except for containing occurrences of W_i at all the places where P contains free occurrences of X_1. Then*

$$\mathbf{Ps}^\infty(P) = \underset{N\to\infty}{Limit} \sum_{i=1}^{N} (\mathbf{w}^N(W_i)\cdot\mathbf{Ps}^\infty(P_i))$$

if such a limit exists; otherwise $\mathbf{Ps}^\infty(P)$ has no value.

(d2) *Let P be an open sentence of L^∞; let X_1, X_2, $\cdots$, and X_{n+1} be the $n+1$ ($n \geq 1$) individual variables of L^∞ which are free in P; and for each i from 1 on let W_i be the i-th individual constant of L^∞ and P_i be like P except for containing occurrences of W_i at all the places where P contains free occurrences of X_{n+1}. Then*

$$\mathbf{Ps}^\infty(P) = \underset{N\to\infty}{Limit} \sum_{i=1}^{N} (\mathbf{w}^N(W_i)\cdot\mathbf{Ps}^\infty(P_i))$$

if such a limit exists; otherwise $\mathbf{Ps}^\infty(P)$ has no value.

Proof by D13.2.[9]

According to T13.5(c2), $\mathbf{Ps}^N(w \text{ outlives } x)$ equals

$$\sum_{i=1}^{N} (\mathbf{w}^N(W_i)\cdot\mathbf{Ps}^N(w \text{ outlives } W_i)),$$

which, according to T13.5(c1), equals

$$\sum_{i=1}^{N} (\mathbf{w}^N(W_i)\cdot \sum_{j=1}^{N} (\mathbf{w}^N(W_j)\cdot\mathbf{Ps}^N(W_j \text{ outlives } W_i))),$$

where in both cases W_i is, for each i from 1 to N, the i-th individual constant of L^N and where in the second case W_j is, for each j from 1 to N, the j-th individual constant of L^N.

T13.6. *For each N from 1 on let*

(1) $$\mathbf{w}^N(W_1) = \mathbf{w}^N(W_2),$$

where W_1 and W_2 are any two individual constants of L^N,

(2) $$\mathbf{w}^N(W_1W_2) = \mathbf{w}^N(W_3W_4),$$

where W_1W_2 and W_3W_4 are any two pairs of individual constants of L^N, and so on.

(a) *Let P be a closed sentence of L^N. Then $\mathbf{Ps}^N(P) = 1$ if P is true in L^N, $\mathbf{Ps}^N(P) = 0$ if P is not true in L^N.*

(b) *Let P be a closed sentence of L^∞. Then* $\mathbf{Ps}^\infty(P) = 1$ *if P is true in L^∞, and* $\mathbf{Ps}^\infty(P) = 0$ *if P is not true in L^∞.*

(c1) *Let P be an open sentence of L^N; let X_1 be the one individual variable of L^N which is free in P; and for each i from 1 to N let P_i be like P except for containing occurrences of the i-th individual constant of L^N at all the places where P contains free occurrences of X_1. Then*

$$\mathbf{Ps}^N(P) = \sum_{i=1}^{N} (\mathbf{Ps}^N(P_i)/N).$$

(c2) *Let P be an open sentence of L^N; let X_1, X_2, $\cdots$, and X_{n+1} be the $n + 1$ $(n \geq 1)$ individual variables of L^N which are free in P; and for each i from 1 to N let P_i be like P except for containing occurrences of the i-th individual constant of L^N at all the places where P contains free occurrences of X_{n+1}. Then*

$$\mathbf{Ps}^N(P) = \sum_{i=1}^{N} (\mathbf{Ps}^N(P_i)/N).$$

(d1) *Let P be an open sentence of L^∞; let X_1 be the one individual variable of L^∞ which is free in P; and for each i from 1 on let P_i be like P except for containing occurrences of the i-th individual constant of L^∞ at all the places where P contains free occurrences of X_1. Then*

$$\mathbf{Ps}^\infty(P) = \mathop{Limit}_{N\to\infty} \sum_{i=1}^{N} (\mathbf{Ps}^\infty(P_i)/N)$$

if such a limit exists; otherwise $\mathbf{Ps}^\infty(P)$ *has no value.*

(d2) *Let P be an open sentence of L^∞; let X_1, X_2, $\cdots$, and X_{n+1} be the $n + 1$ $(n \geq 1)$ individual variables of L^∞ which are free in P; and for each i from 1 on let P_i be like P except for containing occurrences of the i-th individual constant of L^∞ at all the places where P contains free occurrences of X_{n+1}. Then*

$$\mathbf{Ps}^\infty(P) = \mathop{Limit}_{N\to\infty} \sum_{i=1}^{N} (\mathbf{Ps}^\infty(P_i)/N)$$

if such a limit exists; otherwise $\mathbf{Ps}^\infty(P)$ *has no value.*

Proof by D13.2.

According to T13.6(c2), $\mathbf{Ps}^N(w \text{ outlives } x)$ equals

$$\sum_{i=1}^{N} (\mathbf{Ps}^N(w \text{ outlives } W_i)/N),$$

which, according to T13.6(c1), equals

$$\sum_{i=1}^{N}\sum_{j=1}^{N} (\mathbf{Ps}^N(W_j \text{ outlives } W_i)/N^2).^{10}$$

14. STATISTICAL PROBABILITIES AS TRUTH-VALUES

Since the statistical probability allotted in D13.2(a)–(b) to a closed sentence of L coincides, by virtue of D3.4, with the truth-value of that sentence, D13.2 yields the following corollaries:

T14.1. (a)–(b) *Let P be a closed sentence of L. Then*

$$\mathbf{Ps}(P) = \mathbf{Tv}(P).$$

(c) *Let P be an open sentence of L^N; let X_1, X_2, $\cdots$, and X_n be the n ($n \geq 1$) individual variables of L^N which are free in P; and for each i from 1 to N^n let $W_{i_1}W_{i_2} \cdots W_{i_n}$ be the i-th of the sequences made up of any n individual constants of L^N and P_i be like P except for containing occurrences of W_{i_1}, W_{i_2}, $\cdots$, and W_{i_n} at all the places where P contains free occurrences of X_1, X_2, $\cdots$, and X_n, respectively. Then*

$$\mathbf{Ps}^N(P) = \sum_{i=1}^{N^n} (\mathbf{w}^N(W_{i_1}W_{i_2} \cdots W_{i_n})\cdot\mathbf{Tv}^N(P_i)).$$

(d) *Let P be an open sentence of L^∞; let X_1, X_2, $\cdots$, and X_n be the n ($n \geq 1$) individual variables of L^∞ which are free in P; and for each N from 1 on and each i from 1 to N^n let $W_{i_1}W_{i_2} \cdots W_{i_n}$ and P_i be as in* (c). *Then*

$$\mathbf{Ps}^\infty(P) = \underset{N\to\infty}{Limit} \sum_{i=1}^{N^n} (\mathbf{w}^N(W_{i_1}W_{i_2} \cdots W_{i_n})\cdot\mathbf{Tv}^\infty(P_i))$$

if such a limit exists; otherwise $\mathbf{Ps}^\infty(P)$ *has no value.*

Proof by D13.2 and D3.4.

T14.2. *For each N from 1 on let*

(1) $$\mathbf{w}^N(W_1) = \mathbf{w}^N(W_2),$$

where W_1 and W_2 are any two individual constants of L^N,

(2) $$\mathbf{w}^N(W_1W_2) = \mathbf{w}^N(W_3W_4),$$

where W_1W_2 and W_3W_4 are any two pairs of individual constants of L^N, and so on.

(a)–(b) *Let P be a closed sentence of L. Then*

$$\mathbf{Ps}(P) = \mathbf{Tv}(P).$$

(c) *Let P be an open sentence of L^N; let n ($n \geq 1$) be the number of individual variables of L^N which are free in P; and let P_1^*, P_2^*, $\cdots$, and $P_{N^n}^*$ be in some arbitrary order the N^n instances of P in L^N. Then*

$$\mathbf{Ps}^N(P) = \sum_{i=1}^{N^n} (\mathbf{Tv}^N(P_i^*)/N^n).$$

(d) *Let P be an open sentence of L^∞; let n ($n \geq 1$) be the number of individual variables of L^∞ which are free in P; and for each N from 1 on let P_1^*, P_2^*, $\cdots$, and $P_{N^n}^*$ be in some arbitrary order the N^n instances of P in L^N. Then*

$$\mathbf{Ps}^\infty (P) = \underset{N\to\infty}{Limit} \sum_{i=1}^{N^n} (\mathbf{Tv}^\infty (P_i^*)/N^n)$$

if such a limit exists; otherwise $\mathbf{Ps}^\infty$ *(P) has no value.*

Proof by D13.2, D3.4, and D2.6(b).

The two corollaries are of interest. Consider first T14.1(a)–(b), where P is presumed to be a closed sentence of L. By D2.6(a) P is the one and only instance of P in L. The statistical probability in L of a closed sentence P of L thus proves by T14.1(a)–(b) to be the truth-value in L of the one and only instance of P in L. Consider next T14.1(c), where P is presumed to be an open sentence of L^N. By D2.6(b) the sentences P_1, P_2, $\cdots$, and P_{N^n} described in the preamble to T14.1(c) are the N^n instances of P in L^N. The statistical probability in L^N of an open sentence P of L^N thus proves by T14.1(c) to be a weighted sum of the truth-values in L^N of the N^n instances of P in L^N.[11] Consider finally T14.1(d), where P is presumed to be an open sentence of L^∞. By D2.6(b) the sentences P_1, P_2, $\cdots$, and P_{N^n} described in the preamble to T14.1(d) are, for each N from 1 on, the N^n instances of P in L^N. The statistical probability in L^∞ of an open sentence P of L^∞ thus proves by T14.1(d) to be the limit (if any) for increasing N of a weighted sum of the truth-values in L^∞ of the N^n instances of P in L^N (or, if you care to phrase it so, of the first N^n instances of P in L^∞).

A more striking result yet is in store when, as in T14.2, any two sequences of n ($n \geq 1$) individual constants of $L^N (N = 1, 2, 3, \cdots)$

are allotted the same weight (namely, 1 over N^n) in L^N. The statistical probability in L of a closed sentence P of L then proves to be (as before) the average truth-value in L of the one and only instance of P in L; the statistical probability in L^N of an open sentence P of L^N proves to be the average truth-value in L^N of the N^n instances of P in L^N; and the statistical probability in L^∞ of an open sentence P of L^∞ proves to be the limit (if any) for increasing N of the average truth-value in L^∞ of the N^n instances of P in L^N (or, if you care to phrase it so, of the first N^n instances of P in L^∞).

I may now have gone some way towards bringing statistical probabilities into line with inductive probabilities. I have shown that the statistical probabilities currently allotted to sets can be turned into sentence-theoretic measurements. I have also shown that the sentence-theoretic measurements in question are truth-values, or sums of truth-values, or limits of sums of truth-values (that is, truth-values in the traditional sense of the word or in a generalized sense of the word). In Chapter 4 I shall show that the inductive probabilities currently allotted to sentences qualify as estimates of statistical probabilities. It will then appear that all probabilities may be treated as measurements on sentences, that statistical probabilities may be treated as truth-values, and that inductive probabilities may be treated as estimates of truth-values.

15. THE RELATIVE FREQUENCY ALLOTMENT

The simplest allotment of weights endorsed in D13.1 is the relative frequency one, which we have already met in T13.6 and T14.2; historically speaking, it is also the most important one.[12] I first show that the absolute probability in L of a sentence of L, as reckoned under the relative frequency allotment, meets suitable analogues of restrictions (b1)–(b3) on page 34 (T15.1–6). I then show that the conditional probability in L of a pair of sentences of L, as reckoned under the same allotment, meets suitable analogues of restrictions (c1)–(c4) on page 36 (T15.7–11). The reader may verify on his own that theorems T15.1–11 hold as well under the remaining allotments of weights endorsed in D13.1. To condense matters, I attach "**(Rf)**" to the call number of every theorem meant to be prefaced by the clause: *For each N from* 1 *on let* (1) $\mathbf{w}^N(W_1) = \mathbf{w}^N(W_2)$, *where* W_1 *and* W_2 *are any two individual constants of* L^N, (2) $\mathbf{w}^N(W_1W_2) = \mathbf{w}^N(W_3W_4)$, *where* W_1W_2 *and* W_3W_4 *are any two pairs of individual constants of* L^N, *and so on;* I also presume that every mention of $\mathbf{Ps}^\infty$ (P) (or of $\mathbf{Ps}(P)$, where P is a sentence of L^∞) implicitly carries the qualification: *if* $\mathbf{Ps}^\infty$ (P) (*or* $\mathbf{Ps}(P)$) *has a value;* I finally presume that every mention of a limit implicitly carries the qualification: *if such a limit exists.*[13]

According to T15.1, Ps(P), when reckoned under the relative frequency allotment, meets restriction (b1) on page 34 and, hence, is non-negative.

T15.1(Rf). *$0 \leq Ps(P)$.*

Proof: (a)–(b) P is a closed sentence of L. $0 \leq \mathbf{Tv}(P)$ by D3.4. Hence T15.1 by T14.2(a)–(b). (c) P is an open sentence of L^N. Let n ($n \geq 1$) be the number of individual variables of L^N which are free in P, and let $P_1^*, P_2^*, \cdots$, and $P_{N^n}^*$ be in some arbitrary order the N^n instances of P in L^N. For each i from 1 to N^n, $0 \leq \mathbf{Tv}^N(P_i^*)$ by D3.4, D2.6(b), and D2.4. Hence T15.1 by T14.2(c). (d) P is an open sentence of L^∞. Same proof as in (c), but using T14.2(d) instead of T14.2(c).[14]

According to T15.2, $\mathbf{Ps}(P)$, when reckoned under the relative frequency allotment, meets restriction (b2) on page 34 and, hence, is what mathematicians call normed.[15]

T15.2(Rf). *If every instance of P in L is true in L, then* $\mathbf{Ps}(P) = 1$.

Proof: (a)–(b) P is a closed sentence of L. $\mathbf{Tv}(P) = 1$ by D3.4(a) and D2.6(a). Hence T15.2 by T14.2(a)–(b). (c) P is an open sentence of L^N. Let n ($n \geq 1$) be the number of individual variables of L^N which are free in P, and let $P_1^*, P_2^*, \cdots$, and $P_{N^n}^*$ be in some arbitrary order the N^n instances of P in L^N. If every instance of P in L^N is true in L^N, then, for each i from 1 to N^n, $\mathbf{Tv}(P_i^*) = 1$ by D3.4(a), D2.6(b), and D2.4. Hence T15.2 by T14.2(c). (d) P is an open sentence of L^∞. Same proof as in (c), but using T14.2(d) instead of T14.2(c).

Theorem T15.3 serves as a lemma to theorems T15.4 and T15.5, which serve in turn as lemmas to T15.6.

T15.3(Rf). (a) *Let P be an open sentence of L^N and, for each i from 1 to N, let P_i be like P except for containing occurrences of the i-th individual constant of L^N at all the places where P contains free occurrences of some individual variable of L^N. Then*

$$\mathbf{Ps}^N(P) = \sum_{i=1}^{N} \frac{\mathbf{Ps}^N(P_i)}{N}.$$

(b) *Let P be an open sentence of L^∞ and, for each i from 1 on, let P_i be like P except for containing occurrences of the i-th individual constant of L^∞ at all the places where P contains free occurrences of some individual variable of L^∞. Then*

$$\mathbf{Ps}^\infty (P) = \mathit{Limit}_{N\to\infty} \sum_{i=1}^{N} \frac{\mathbf{Ps}^\infty (P_i)}{N}.$$

Proof by D13.2.

Theorems T15.4 and T15.5 are proved by mathematical induction, a device which may call here for a word of explanation. Consider first T15.5. I wish to show that $\mathbf{Ps}((P \,\&\, {\sim} Q) \vee ({\sim} P \,\&\, Q))$ is equal to $\mathbf{Ps}(P \,\&\, {\sim} Q) + \mathbf{Ps}({\sim} P \,\&\, Q)$. To establish this, I first show (step 1) that T15.5 holds when n, the number of individual variables of L free in $(P \,\&\, {\sim} Q) \vee ({\sim} P \,\&\, Q)$, is equal to 0; then, I show (step 2) that if T15.5 holds for any given n (a hypothesis called the hypothesis of induction), T15.5 must also hold for $n + 1$. It follows from step 1 and step 2 that T15.5 holds for every n. T15.5 holds indeed by step 1 for 0, hence by step 2 for 1, hence by step 2 again for 2, and so on ad infinitum. Or consider T15.4. I wish to show that if every instance of $P \equiv Q$ in L is true in L, then $\mathbf{Ps}(P) = \mathbf{Ps}(Q)$. To establish this, I first show (step 1) that T15.4 holds in case $\max(p,q)$ equals 0, where p is the number of individual variables of L free in P, q is the number of individual variables of L free in Q, and $\max(p,q)$ is the larger one of the two numbers p and q; then, I show (step 2) that if T15.4 holds for any given $\max(p,q)$ (a hypothesis called again the hypothesis of induction), T15.4 must also hold for $\max(p,q) + 1$. By the same reasoning as above, it follows from step 1 and step 2 that T15.4 holds for every $\max(p,q)$. Mathematical induction, incidentally, is one of the commonest ways of proving theorems in so-called metalogic.

T15.4(Rf). *If every instance of* $P \equiv Q$ *in* L *is true in* L, *then* $\mathbf{Ps}(P) = \mathbf{Ps}(Q)$.[16]

Proof by mathematical induction on $\max(p,q)$, where p is the number of individual variables of L free in P and q is the number of individual variables of L free in Q.

STEP 1: $\text{Max}(p,q) = 0$. Then by D2.4 P and Q are two closed sentences of L and hence by D2.9, D2.7, and D2.3–4 $P \equiv Q$ is a closed sentence of L. But if $P \equiv Q$ is a closed sentence of L and every instance of $P \equiv Q$ in L is true in L, then by D2.6(a) $P \equiv Q$ is true in L and hence by D2.9, D2.7, and D3.2 P is true in L if and only if Q is true in L. Hence T15.4 by T14.2(a)–(b).

STEP 2:

PART ONE: (a) If both p and q are larger than 0 and at least one individual variable of L which is free in P is also free in Q, then for each i from 1 to N or from 1 on let P_i and Q_i respectively be like P and Q except for containing occurrences of the i-th individual constant of L at all the places where P and Q contain free occurrences of W, W being in alphabetical order the first individual variable of L which is free in both P and Q. (b) If both p and q are larger than 0 but no individual variable of L which is free in P is also free in Q, then for each i from 1 to N or from 1 on let P_i and Q_i respectively be like P and Q except for containing occurrences of the i-th individual constant of L at all the places where P contains free occurrences of W and Q contains free occurrences of X, W being in alphabetical order the first individual variable of L which is free in P and X being in alphabetical order the first individual variable of L which is free in Q. (c) If p is larger than 0 but q is equal to 0, then for each i from 1 to N or from 1 on let Q_i be Q and let P_i be like P except for containing occurrences of the i-th individual constant of L at all the places where P contains free occurrences of W, W being in alphabetical order the first individual variable of L which is free in P. (d) If q is larger than 0 but p is equal to 0, then for each i from 1 to N or from 1 on let P_i be P and let Q_i be like Q except for containing occurrences of the i-th individual constant of L at all the places where Q contains free occurrences of W, W being in alphabetical order the first individual variable of L which is free in Q.

PART TWO: If every instance of $P \equiv Q$ in L is true in L, then by D2.6(b) every instance of $P_i \equiv Q_i$ in L is true in L, and hence by the hypothesis of induction $\mathbf{Ps}(P_i) = \mathbf{Ps}(Q_i)$ for each i from 1 to N or from 1 on. Hence T15.4 by T14.2(a)–(b) and/or T15.3.

To illustrate the workings of the above proof, let 'c_1', 'c_2', and 'c_3' be short for the three individual constants of L^3, let 'G' and 'G′' be short for two predicates of L^3, let $P \equiv Q$ be '$G(w,x) \equiv G'(y)$', and let every instance of '$G(w,x) \equiv G'(y)$' in L^3 be true in L^3. (1) Consider first '$G(c_1,c_1) \equiv G'(c_1)$,' '$G(c_2,c_1) \equiv G'(c_1)$,' '$G(c_3,c_1) \equiv G'(c_1)$', '$G(c_1,c_2) \equiv G'(c_2)$', '$G(c_2,c_2) \equiv G'(c_2)$', '$G(c_3,c_2) \equiv G'(c_2)$', '$G(c_1,c_3) \equiv G'(c_3)$', '$G(c_2,c_3) \equiv G'(c_3)$', and '$G(c_3,c_3) \equiv G'(c_3)$'. $\mathrm{Max}(p,q)$ is equal in all nine cases to 0. Hence by virtue of step 1

$$\mathbf{Ps}^3(G(c_1,c_1)) = \mathbf{Ps}^3(G'(c_1))$$
$$\mathbf{Ps}^3(G(c_2,c_1)) = \mathbf{Ps}^3(G'(c_1))$$
$$\mathbf{Ps}^3(G(c_3,c_1)) = \mathbf{Ps}^3(G'(c_1))$$
$$\mathbf{Ps}^3(G(c_1,c_2)) = \mathbf{Ps}^3(G'(c_2))$$
$$\mathbf{Ps}^3(G(c_2,c_2)) = \mathbf{Ps}^3(G'(c_2))$$
$$\mathbf{Ps}^3(G(c_3,c_2)) = \mathbf{Ps}^3(G'(c_2))$$
$$\mathbf{Ps}^3(G(c_1,c_3)) = \mathbf{Ps}^3(G'(c_3))$$
$$\mathbf{Ps}^3(G(c_2,c_3)) = \mathbf{Ps}^3(G'(c_3))$$

and

$$\mathbf{Ps}^3(G(c_3,c_3)) = \mathbf{Ps}^3(G'(c_3)).$$

(2.1) Consider next '$G(w,c_1) \equiv G'(c_1)$'. $\mathrm{Max}(p,q)$ is equal here to 1. Hence by virtue of step 2 and the first three equalities under (1)

$$\mathbf{Ps}^3(G(w,c_1)) = \mathbf{Ps}^3(G'(c_1)).$$

(2.2) Consider next '$G(w,c_2) \equiv G'(c_2)$'. $\mathrm{Max}(p,q)$ is equal here again to 1. Hence by virtue of step 2 and the next three equalities under (1)

$$\mathbf{Ps}^3(G(w,c_2)) = \mathbf{Ps}^3(G'(c_2)).$$

(2.3) Consider next '$G(w,c_3) \equiv G'(c_3)$'. $\mathrm{Max}(p,q)$ is equal here again to 1. Hence by virtue of step 2 and the last three equalities under (1)

$$\mathbf{Ps}^3(G(w,c_3)) = \mathbf{Ps}^3(G'(c_3)).$$

(3) Consider finally '$G(w,x) \equiv G'(y)$'. $\mathrm{Max}(p,q)$ is equal here to 2. Hence by virtue of step 2 and the three equalities under (2.1)–(2.3)

$$\mathbf{Ps}^3(G(w,x)) = \mathbf{Ps}^3(G'(y)).$$

T15.5(Rf). $\mathbf{Ps}((P \mathbin{\&} \sim Q) \vee (\sim P \mathbin{\&} Q))$

$$= \mathbf{Ps}(P \mathbin{\&} \sim Q) + \mathbf{Ps}(\sim P \mathbin{\&} Q).$$

Proof by mathematical induction on the number n ($n \geq 0$) of individual variables of L which are free in $(P \mathbin{\&} \sim Q) \vee (\sim P \mathbin{\&} Q)$ and hence, by D2.7–8 and D2.3, in $P \mathbin{\&} \sim Q$ and $\sim P \mathbin{\&} Q$.

STEP 1: $n = 0$. Then by D2.4 $(P \mathbin{\&} \sim Q) \vee (\sim P \mathbin{\&} Q)$, $P \mathbin{\&} \sim Q$, and $\sim P \mathbin{\&} Q$ are closed sentences of L. By D2.7 and D3.2 $P \mathbin{\&} \sim Q$ is true in L if and only if P is true in L and Q is not true in L, $\sim P \mathbin{\&} Q$ is true in L if and only if P is not true in L and Q is true in L, and hence by D2.8 $(P \mathbin{\&} \sim Q) \vee (\sim P \mathbin{\&} Q)$ is true in L if and only if exactly one of $P \mathbin{\&} \sim Q$ and $\sim P \mathbin{\&} Q$ is true in L. Hence T15.5 by T14.2(a)–(b).

STEP 2: For each i from 1 to N or from 1 on let $((P \mathbin{\&} \sim Q) \vee (\sim P \mathbin{\&} Q))_i$, $(P \mathbin{\&} \sim Q)_i$, and $(\sim P \mathbin{\&} Q)_i$ respectively be like $(P \mathbin{\&} \sim Q) \vee (\sim P \mathbin{\&} Q)$, $P \mathbin{\&} \sim Q$, and $\sim P \mathbin{\&} Q$ except for containing occurrences of the i-th individual constant of L at all the places where $(P \mathbin{\&} \sim Q) \vee (\sim P \mathbin{\&} Q)$, $P \mathbin{\&} \sim Q$, and $\sim P \mathbin{\&} Q$ contain free occurrences of some individual variable of L. By the hypothesis of induction $\mathbf{Ps}(((P \mathbin{\&} \sim Q) \vee (\sim P \mathbin{\&} Q))_i)$ equals $\mathbf{Ps}((P \mathbin{\&} \sim Q)_i) + \mathbf{Ps}((\sim P \mathbin{\&} Q)_i)$. Hence T15.5 by T15.3.

According to T15.6, $\mathbf{Ps}(P)$, when reckoned under the relative frequency allotment, meets the analogue here of restriction (b3) on page 34 and hence is what mathematicians call finitely additive.[17]

T15.6(Rf). *If every instance of* $\sim (P \mathbin{\&} Q)$ *in* L *is true in* L, *then* $\mathbf{Ps}(P \vee Q) = \mathbf{Ps}(P) + \mathbf{Ps}(Q)$.

Proof: If every instance of $\sim (P \mathbin{\&} Q)$ in L is true in L, then by D2.7–9, D3.2, and D2.6 every instance of $(P \vee Q) \equiv ((P \mathbin{\&} \sim Q) \vee (\sim P \mathbin{\&} Q))$, $P \equiv (P \mathbin{\&} \sim Q)$, and $Q \equiv (\sim P \mathbin{\&} Q)$ in L is true in L, hence by T15.4 $\mathbf{Ps}(P \vee Q)$ equals $\mathbf{Ps}((P \mathbin{\&} \sim Q) \vee (\sim P \mathbin{\&} Q))$, $\mathbf{Ps}(P)$ equals $\mathbf{Ps}(P \mathbin{\&} \sim Q)$, and $\mathbf{Ps}(Q)$ equals $\mathbf{Ps}(\sim P \mathbin{\&} Q)$, and hence by T15.5 $\mathbf{Ps}(P \vee Q)$ equals $\mathbf{Ps}(P) + \mathbf{Ps}(Q)$.

According to T15.7, T15.8, T15.10, and T15.11, $\mathbf{Ps}(P, Q)$, when reckoned under the relative frequency allotment, meets the analogues here of restrictions (c1)–(c4) on page 36. T15.9 serves as a lemma to T15.10–11.

T15.7(Rf). $0 \leq \mathbf{Ps}(P, Q)$.

Proof: Case 1: $\mathbf{Ps}(Q) \neq 0$. Then T15.7 by T15.1 and D13.3(a). Case 2: $\mathbf{Ps}(Q) = 0$. Then T15.7 by D13.3(b).

T15.8(Rf). $\mathbf{Ps}(P, P) = 1$.

Proof: Case 1: $\mathbf{Ps}(P) \neq 0$. By D2.9, D2.7, D3.2, and D2.6 every instance of $(P \,\&\, P) \equiv P$ in L is true in L, and hence by T15.4 $\mathbf{Ps}(P \,\&\, P)$ equals $\mathbf{Ps}(P)$. Hence T15.8 by D13.3(a). Case 2: $\mathbf{Ps}(P) = 0$. Then T15.8 by D13.3(b).

T15.9(Rf). $\mathbf{Ps}(P) = \mathbf{Ps}(Q \,\&\, P) + \mathbf{Ps}(\sim Q \,\&\, P)$.

Proof: By D2.7, D3.2, and D2.6 every instance of $\sim((Q \,\&\, P) \,\&\, (\sim Q \,\&\, P))$ in L is true in L, and hence by T15.6 $\mathbf{Ps}((Q \,\&\, P) \vee (\sim Q \,\&\, P))$ equals $\mathbf{Ps}(Q \,\&\, P) + \mathbf{Ps}(\sim Q \,\&\, P)$. By D2.7–9, D3.2, and D2.6, on the other hand, every instance of $P \equiv ((Q \,\&\, P) \vee (\sim Q \,\&\, P))$ in L is true in L, and hence by T15.4 $\mathbf{Ps}(P)$ equals $\mathbf{Ps}((Q \,\&\, P) \vee (\sim Q \,\&\, P))$. Hence T15.9.

T15.10(Rf). $\mathbf{Ps}(P \,\&\, Q, R) = \mathbf{Ps}(P, Q \,\&\, R) \cdot \mathbf{Ps}(Q, R)$.

Proof: Case 1: $\mathbf{Ps}(R) \neq 0$. Subcase 1.1: $\mathbf{Ps}(Q \,\&\, R) \neq 0$. By D2.9, D2.7, D3.2, and D2.6 every instance of $((P \,\&\, Q) \,\&\, R) \equiv (P \,\&\, (Q \,\&\, R))$ in L is true in L, and hence by T15.4 $\mathbf{Ps}((P \,\&\, Q) \,\&\, R)$ equals $\mathbf{Ps}(P \,\&\, (Q \,\&\, R))$. Hence T15.10 by D13.3(a). Subcase 1.2: $\mathbf{Ps}(Q \,\&\, R) = 0$. Then by T15.9 and T15.1 $\mathbf{Ps}(P \,\&\, (Q \,\&\, R))$ equals 0. But by the same reasoning as in subcase 1.1 $\mathbf{Ps}((P \,\&\, Q) \,\&\, R)$ equals $\mathbf{Ps}(P \,\&\, (Q \,\&\, R))$. Hence T15.10 by D13.3. Case 2: $\mathbf{Ps}(R) = 0$. Then by T15.9 and T15.1 $\mathbf{Ps}(Q \,\&\, R)$ equals 0. Hence T15.10 by D13.3(b).

T15.11(Rf). *If* $\mathbf{Ps}(Q) \neq 0$, *then* $\mathbf{Ps}(\sim P, Q) = 1 - \mathbf{Ps}(P, Q)$.

Proof by T15.9 and D13.3(a).

I append to the section two invariance theorems; they are analogues for $\mathbf{Ps}(P, Q)$ of T15.4.

T15.12(Rf). *If every instance of* $P \equiv Q$ *in* L *is true in* L, *then* $\mathbf{Ps}(P, R) = \mathbf{Ps}(Q, R)$.

Proof: Case 1: $\mathbf{Ps}(R) \neq 0$. If every instance of $P \equiv Q$ in L is true in L, then by D2.9, D2.7, D3.2, and D2.6 every instance of $(P \,\&\, R) \equiv (Q \,\&\, R)$ in L is true in L, and hence by T15.4 $\mathbf{Ps}(P \,\&\, R)$

equals **Ps**(Q & R). Hence T15.12 by D13.3(a). CASE 2: **Ps**(R) $= 0$. Then T15.12 by D13.3(b).

T15.13(Rf). *If every instance of* $Q \equiv R$ *in* L *is true in* L, *then* **Ps**(P, Q) $=$ **Ps**(P, R).

Proof: If every instance of $Q \equiv R$ in L is true in L, then by T15.4 **Ps**(Q) equals **Ps**(R). Hence only two cases arise. CASE 1: **Ps**(Q) $\neq 0$ and **Ps**(R) $\neq 0$. If every instance of $Q \equiv R$ in L is true in L, then by D2.9, D2.7, D3.2, and D2.6 every instance of (P & Q) $\equiv$ (P & R) in L is true in L, and hence by T15.4 **Ps**(P & Q) equals **Ps**(P & R). Hence T15.13 by D13.3(a). CASE 2: **Ps**(Q) $= 0$ and **Ps**(R) $= 0$. Then T15.13 by D13.3(b).

16. THE STATISTICAL PROBABILITY OF CLOSED SENTENCES

Consider the two closed sentences 'John Doe owns a station wagon' and 'John Doe is a suburbanite.' Three questions immediately arise: What is the statistical probability of 'John Doe owns a station wagon'? What is the statistical probability of 'John Doe is a suburbanite'? And what is the statistical probability of 'John Doe owns a station wagon' given 'John Doe is a suburbanite'? By virtue of D13.2(a) and D13.3, the answer to the first question is 1 if John owns a station wagon, 0 if he doesn't; the answer to the second is 1 if John is a suburbanite, 0 if he isn't; and the answer to the third is 1 if John is a suburbanite and owns a station wagon, 0 if John is a suburbanite and doesn't own a station wagon, and 1 again if John isn't a suburbanite.

Idle though the answers may sound, they are in my opinion the only fitting ones. As I remarked in the Preface, statistical probabilities betoken some indefiniteness in the sentences whose probability we assess, indefiniteness as to what their exact subject matter may be. But there is nothing indefinite in a closed sentence. Its statistical probability or that of a pair of closed sentences should therefore be 1 or 0.

Idle though the answers may sound, they are also the only ones which jibe with certain intuitive requisites of probability theory. Imagine, for example, that a closed sentence P of L were not allotted any probability. The logically equivalent sentence $W = W \supset P$, where W is an individual variable of L, would, nevertheless, be allotted one by virtue of D13.2(c) or D13.2(d), and a first requisite of probability theory, namely, that a sentence be allotted a probability if any sentence logically equivalent to it is allotted one, would be violated.[18] Imagine, on the other hand, that P were allotted some probability other than 1 or 0. By virtue of D3.2 and either D13.2(c) or D13.2(d), the logically equivalent sentence $W = W \supset P$, where W is as above, would, nevertheless, be allotted 1 or 0 as its probability, and a second requisite of probability theory, namely, that any two logically equivalent sentences be allotted the same probability if they are allotted any at all, would be violated.[19]

The two requisites I just mentioned fare very badly at Kemeny's hands. Given a (finite) set $\vee_P$ of so-called logical possibilities, an allotment of weights to the members of $\vee_P$, and a closed sentence P of apparently any form whatever, Kemeny takes the probability of P to be the combined weights of all the members of $\vee_P$ which are not—so to speak—precluded by P (1). Given a (finite) set $\vee_I$ of individuals, an allotment of weights to the members of $\vee_I$, and an open sentence P of the form $F(W) = X$, where F is a one-place functor, W an individual constant, and X an individual variable, Kemeny next takes as the probability of P the combined weights of all the members of $\vee_I$ which satisfy P when assigned to X (2).[20]

By limiting himself to open sentences of the form $F(W) = X$, Kemeny violates the first of our two requisites since he cannot allot probabilities to the logically equivalent sentences $X = X \supset F(W) = X$, $(X = X \,\&\, Y = Y) \supset F(W) = X$, where Y is an individual variable distinct from X, and so on. The oversight, to be sure, is easily remedied: we need only let P in (2) be any open sentence we please, allot weights to the pairs of members of $\vee_I$, triples of members of $\vee_I$, and so on, as well as to the members of $\vee_I$, and, presuming $X_1, X_2, \cdots$, and X_n ($n \geq 1$) to be all the individual variables which are free in P, take as the probability of P the combined weights of all the sequences of members of $\vee_I$ which satisfy P when the members of $\vee_I$ of which the said sequences are made up have been assigned to $X_1, X_2, \cdots$, and X_n. Note, however, that with (2) thus amended, $X = X \supset F(W) = Y$, where W and Y are two individual constants and X is an individual variable, comes to be allotted 1 or 0 as its probability,[21] whereas the closed sentence $F(W) = Y$, a sentence logically equivalent to $X = X \supset F(W) = Y$, can be allotted under (2) some probability other than 1 or 0. Kemeny can therefore not come to terms with our first requisite without falling out with the second, an awkward situation.[22]

The probabilities which Kemeny allots to closed sentences have an inductive flavor of their own and hence will be reconsidered in Chapter 4. Probabilities with a (possibly) more statistical flavor can, however, be allotted to the closed sentences of L in the following manner.

We may not know, for example, whether the closed sentence 'John Doe owns a station wagon' of a given language L^N is true in L^N. We may, however, know how many members of the universe of discourse of L^N happen to own station wagons, and hence be able to reckon the statistical probability in L^N of 'w owns a station wagon,' an open sentence of which 'John Doe owns a station wagon' is an instance in L^N. We may not know either whether the closed sentence 'John Doe is a suburbanite' of L^N is true in L^N. We may, however, know how many members of the universe of discourse of L^N happen to be suburbanites, and hence be able to reckon the statistical probability in L^N of 'w is a suburbanite,' an open sentence of which 'John Doe is a suburbanite' is an instance in L^N. We may not know, finally, whether the closed sentence 'John Doe owns a station wagon & John Doe is a suburbanite' of L^N is true in L^N. We may know, however, how many members of the universe of discourse of L^N happen to own station wagons and be suburbanites, and hence be able to reckon the statistical probability in L^N of 'w owns a station wagon & w is a suburbanite,' an open sentence of which 'John Doe owns a station wagon & John Doe is a suburbanite' is an instance in L^N.[23] We might accordingly take the statistical probability in L^N of 'John Doe owns a station wagon' to be that of 'w owns a station wagon,' the statistical probability in L^N of 'John Doe is a suburbanite' to be that of 'w is a suburbanite,' and the statistical probability in L^N of 'John Doe owns a station wagon' given 'John Doe is a suburbanite' to be that of 'w owns a station wagon' given 'w is a suburbanite.'

Loaning out to closed sentences the probabilities allotted by D13.2(c)–(d) to open ones may upon occasion prove to be a delicate matter. Which probability should we loan out, for example, to 'Jane Doe can't stand John Doe,' a closed sentence which, unlike 'John Doe is a suburbanite,' contains occurrences of more than one individual constant? [24] Or which probability should we loan out to '$(\forall w)$(Is a metal$(w) \supset$ Is malleable(w)),' a closed sentence which, unlike 'John Doe is a suburbanite,' does not contain occurrences of any individual constant? The two difficulties can be met in various ways, of which the following seems as good as any.

Let P be a closed sentence of L.

CASE 1: One or more individual constants of L occur in P. Assuming the individual constants in question to be (in alphabetical order) W_1, $W_2, \cdots$, and W_n, assuming also the first n individual variables of L which are foreign to P to be (in alphabetical order) $X_1, X_2, \cdots$, and X_n, the reader might first appoint as the individual-constant-free mate of P in L the result of substituting X_1 for W_1, X_2 for $W_2, \cdots$, and X_n for W_n in P. He might then loan out to P the statistical probability of P' in L when $n = 1$, that of P' & $\mathsf{D}(X_1, X_2, \cdots, X_n)$ in L when $n > 1$, P' being in both cases the individual-constant-free mate of P in L. $\mathsf{D}(X_1, X_2, \cdots, X_n)$, by the way, is thrown in with P', when $n > 1$, to insure that P and P & $\mathsf{D}(W_1, W_2, \cdots, W_n)$, two sentences easily shown by D2.9, D2.7, D4.5, and D4.1 to be logically equivalent in L, are loaned out the same probability.

CASE 2: No individual constant of L occurs in P, and hence by D2.2 and D2.4 P is of the form $(\forall W)Q$, where W is free in Q, or has at least one subsequence of the form $(\forall W)Q$, where W is again free in Q. 2.1. Since by D3.2(e) $(\forall W)Q$ is true in L^N if and only if the N instances Q_1^*, Q_2^*, $\cdots$, and Q_N^* of Q in L^N are true in L^N, the reader might loan out to P, when P is a sentence of L^N, the probability loaned out under Case 1 to the result of substituting $(\cdots (Q_1^* \,\&\, Q_2^*) \,\&\, \cdots) \,\&\, Q_N^*$ for the occurrence of $(\forall W)Q$ in P or, if several occurrences of $(\forall W)Q$ there be, for any one occurrence of $(\forall W)Q$ in P. 2.2. Since by D3.2(e) $(\forall W)Q$ is true in L^∞ if and only if the infinitely many instances Q_1^*, Q_2^*, $\cdots$, Q_N^*, $\cdots$, of Q in L^∞ are true in L^∞, the reader might loan out to P, when P is a sentence of L^∞, the limit (if any) for increasing N of the probability loaned out under Case 1 to the result of substituting $(\cdots (Q_1^* \,\&\, Q_2^*) \,\&\, \cdots) \,\&\, Q_N^*$ for an occurrence of $(\forall W)Q$ in P.

By virtue of Case 1, 'Jane Doe can't stand John Doe' would be loaned out in L the probability of 'w can't stand x & D(w, x)'; by virtue of Case 2, '($\forall w$)(Is a metal(w) $\supset$ Is malleable(w))' would be loaned out in L^N the probability of ($\cdots$ ((Is a metal(W_1) $\supset$ Is malleable(W_1)) & (Is a metal(W_2) $\supset$ Is malleable(W_2))) & $\cdots$) & (Is a metal(W_N) $\supset$ Is malleable(W_N)), where W_1, W_2, $\cdots$, and W_N are the N individual constants of L^N; and so on.[25]

I have insisted upon this rival allotment because many writers seem to have something like it in mind when they talk of the probability of a so-called single case.[26] I have also insisted upon it because of the part to be played in Section 17 by the individual-constant-free mate in L of a closed sentence of L.

17. A FIRST LOOK AT INDUCTIVE INFERENCES

Typical of deductive inferences is the one from

All the neighbors of Cabot are Republicans (1)

and

Martin is a neighbor of Cabot (2)

to

Martin is a Republican, (3)

where the conclusion is logically implied by the premises. Typical, on the other hand, of inductive inferences is the one from

Martin is a neighbor of Cabot (4)

and

Martin is a Republican (5)

to

All the neighbors of Cabot are Republicans, (6)

or, to give examples perhaps more familiar to statisticians, the inference from

Nine tenths of the neighbors of Cabot are Republicans (7)

to

Nine tenths of the neighbors of Cabot who show up at the polls will be Republicans, (8)

or the one from

Nine tenths of the neighbors of Cabot who showed up at the polls were Republicans (9)

to

Nine tenths of the neighbors of Cabot are Republicans, (10)

where the conclusion is not implied in any case by the premises.[27]

Deductive inferences are infallible: (1) and (2), for example, cannot be true without (3) being also true. Inductive ones are not: (4) and (5), for example, can be true without (6) being also true, (7) can be true without (8) being also true, and (9) can be true without (10) being also true. Some inductive inferences, however, are less fallible than others or, one might say, some types of inductive inference are less fallible than others. The type of inference leading from (7) to (8), for example, or the one leading from (9) to (10), would commonly be rated less fallible than the one leading from (4) and (5) to (6). Probabilities may have been devised by classical writers to measure the trust one can place in various types of inductive inference. It is my purpose here, at any rate, to show how statistical probabilities may serve as a gauge of that trust.

Suppose we were given the following: (1) a premise P and a conclusion Q, which, for simplicity's sake, I shall assume to be closed sentences of a sublanguage L^N of L^∞ and to contain occurrences of a single individual constant of L^N, the same constant, say W, in each case; (2) the pair of individual-constant-free mates P' and Q' of P and Q in L^N, where P' and Q' are respectively like P and Q except for containing occurrences of the first individual variable of L^N which is foreign to P and Q, say X, at all the places where P and Q contain occurrences of W; and (3) the N pairs, P_1^* and Q_1^*, P_2^* and Q_2^*, $\cdots$, and P_N^* and Q_N^*, of instances of P' and Q' in L^N. Suppose next we knew the truth-value of P_i^* and that of Q_i^* in L^N for each i from 1 to N. Suppose next that, instead of merely drawing Q from P if P

were true in L^N, we went through all the pairs of instances of P' and Q' in L^N and drew Q_1^* from P_1^* if P_1^* were true in L^N, Q_2^* from P_2^* if P_2^* were true in L^N, $\cdots$, and Q_N^* from P_N^* if P_N^* were true in L^N. Suppose next that, as each conclusion were drawn, we chalked one up for ourselves when the conclusion proved to be true in L^N. Suppose finally that, the last conclusion drawn, we added up the ones (if any) we chalked up for ourselves and divided the resulting sum by the number of conclusions drawn in the process. Assuming, as I do here for simplicity's sake, that the individual constants of L^N are all weighted alike in L^N, we might treat the figure thus arrived at as the coefficient of statistical reliability in L^N, $\mathbf{CRs}^N(Q, P)$ for short, of the type of inference leading from P to Q.

Consider now $\mathbf{Ps}^N(Q', P')$, the statistical probability in L^N of Q' given P'. $\mathbf{Ps}^N(Q', P')$ is equal by D13.3(a) to $\mathbf{Ps}^N(Q' \,\&\, P')/\mathbf{Ps}^N(P')$,[28] which is equal by T15.9 to

$$\mathbf{Ps}^N(Q' \,\&\, P')/(\mathbf{Ps}^N(Q' \,\&\, P') + \mathbf{Ps}^N(\sim Q' \,\&\, P')).$$

But by T14.2(a)–(b), D2.7, and D3.2, $\mathbf{Ps}^N(Q' \,\&\, P')$ is equal to the proportion among all the pairs of instances of P' and Q' in L^N of those in which both sentences are true in L^N; $\mathbf{Ps}^N(\sim Q' \,\&\, P')$ is equal to the proportion among all the pairs of instances of P' and $\sim Q'$ in L^N of those in which both sentences are true in L^N; and, hence,

$$\mathbf{Ps}^N(Q' \,\&\, P')/(\mathbf{Ps}^N(Q' \,\&\, P') + \mathbf{Ps}^N(\sim Q' \,\&\, P'))$$

is equal to the proportion of pairs of instances of P' and Q' in L^N among the various pairs of instances of P' and Q' and of P' and $\sim Q'$ in L^N in which both sentences are true in L^N. $\mathbf{Ps}^N(Q',P')$ is thus equal to $\mathbf{CRs}^N(Q,P)$ for any pair of sentences P and Q of the sort under discussion.

The result I have just obtained also holds for other pairs of closed sentences P and Q of L. Various adjustments may be called for, however.

(1) When P and Q are closed sentences of L^∞ rather than of L^N, we must attend to the infinitely many pairs P_1^* and Q_1^*, P_2^* and Q_2^*, $\cdots$, P_N^* and Q_N^*, $\cdots$, of instances of P' and Q' in L^∞ and take $\mathbf{CRs}^\infty(Q, P)$ to be the limit (if any) for increasing N of (a) the sum of the ones we chalk up for ourselves as we draw Q_1^* from P_1^* if P_1^*

is true in L^∞, Q_2^* from P_2^* if P_2^* is true in L^∞, $\cdots$, and Q_N^* from P_N^* if P_N^* is true in L^∞, to (b) the number of conclusions drawn in the process. $\mathbf{CRs}^\infty(Q, P)$, when thus reckoned, will match $\mathbf{Ps}^\infty(Q', P')$.
(2) When P and Q, though each containing occurrences of a single individual constant of L, do not contain occurrences of the same constant or when either one of P and Q contains occurrences of two or more individual constants, we must

(2.1) expand P into P & $\mathsf{D}(W_1, W_2, \cdots, W_n)$ and Q into Q & $\mathsf{D}(W_1, W_2, \cdots, W_n)$, $W_1, W_2, \cdots$, and W_n being in alphabetical order the n individual constants of L which occur in P or Q or both,

(2.2) attend to the N^n or infinitely many pairs of instances of P' & $\mathsf{D}(X_1, X_2, \cdots, X_n)$ and Q' & $\mathsf{D}(X_1, X_2, \cdots, X_n)$ in L, $X_1, X_2, \cdots$, and X_n being in alphabetical order the first n individual variables of L which are foreign to P and Q, P' being like P except for containing, for each i from 1 to n, occurrences of X_i at all the places where P contains occurrences of W_i, and Q' being like Q except for containing, again for each i from 1 to n, occurrences of X_i at all the places where Q contains occurrences of W_i, and

(2.3) reckon $\mathbf{CRs}(Q, P)$ as I did two paragraphs back or as I just did under (1). $\mathbf{CRs}(Q, P)$, thus reckoned, will match $\mathbf{Ps}(Q'$ & $\mathsf{D}(X_1, X_2, \cdots, X_n)$, P' & $\mathsf{D}(X_1, X_2, \cdots, X_n))$.
(3) When either one of P and Q does not contain occurrences of any individual constant of L, we must first expand the sentence in question—or, if need be, the two sentences—along the lines of Case 2 on pages 88–89 and then proceed as in (2). Further adjustments are still called for when the individual constants of L^N, or pairs of individual constants of L^N, and so on, are not all weighted alike.[29]

I have restricted myself here to inferences with a single premise. Any n ($n \geq 2$) premises $P_1, P_2, \cdots$, and P_n can be lumped, however, into a single premise, namely, $(\cdots(P_1$ & $P_2)$ & $\cdots)$ & P_n. Inferences with two or more premises are thus accounted for as well.

What holds true of sentence-theoretic measurements holds true, mutatis mutandis, of set-theoretic ones, a fact statisticians have often exploited. Suppose, for example, we are given a pair of subsets A and B of a probability set PS. Suppose also, to simplify matters, that equal weights are allotted to the members of PS when PS is finite in size, to the members of appropriate subsets of *PS* when *PS* is denumerably infinite in size and serially ordered. Suppose also that the probability of *A* given *B* is equal, say, to 0.9, or 0.99, or 0.999, and so on. Suppose finally that a given member of *PS* is known to belong to *B*. Statisticians would often trust the member of *PS* in question to belong to *A* as well, on the grounds that 90 per cent, or 99 per cent, or 99.9 per cent, and so on, of the members of *B* would be found, under inspection of the two sets, to belong to *A*.

A few caveats must be entered, however. First, a given premise *P* and conclusion *Q* cannot belong to one language *L* without automatically belonging to infinitely many others, a hitch of sorts, since **CRs**(*Q*, *P*) may vary with the languages in question. Second, whatever language *L* the given *P* and *Q* be considered to belong to, the individual constants, pairs of individual constants, and so on, of that *L* (or of its sublanguages, if *L* happens to be L^{∞}) may be allotted infinitely many different weights, another hitch, since **CRs**(*Q*, *P*) may vary with the weights in question. Third, whatever language *L* the given *P* and *Q* be considered to belong to and whatever weights the individual constants, pairs of individual constants, and so on, of that *L* (or of its sublanguages, if *L* happens to be L^{∞}) be allotted, the value of **CRs**(*Q*, *P*) may be unknown. I shall take up in Section 24 the problem of estimating the value of **CRs**(*Q*, *P*) when that value is unknown.

NOTES

[1] A closed sentence may be true in one language *L* of which it is a sentence and yet be false in another. The probability in *L* of a closed sentence of *L* and, as a result, of an open sentence of *L* may thus vary with *L*. I write '$\mathbf{Ps}^{\infty}(P)$' when I wish to emphasize that *P* is a sentence of L^{∞}, '$\mathbf{Ps}^{N}(P)$' when I wish to emphasize that *P* is a sentence of L^{N}; otherwise, I write '**Ps**(*P*),' as I have just done in the text.

[2] See Kemeny, Mirkil, Snell, and Thompson, *loc. cit.*, pp. 65–68 and 112–114. The reference to Kemeny in the text is an abbreviated reference to all four authors. Kemeny's proposal is to be favored, though, when possibilities (1)–(2) on page 9 are left open.

[3] Under my proposal weights also prove (as they did in Chapter 2) to be probabilities of a sort. $\mathbf{w}^{N}$(Mary), for example, proves to be $\mathbf{Ps}^{N}(w = \text{Mary})$

and, to anticipate the next paragraph, $\mathbf{w}^N$(Mary Peter) proves to be $\mathbf{Ps}^N(w =$ Mary & $x =$ Peter$)$. See T13.4 for a full statement of the matter.

[4] The exact ordering does not matter since addition is commutative. One way of ordering the pairs in question is as follows: (1) Let i, for each i from 1 to N, be the index of the i-th individual constant of L^N; (2) let the sum of the respective indexes of two individual constants W_{i_1} and W_{i_2} of L^N be the index of the pair $W_{i_1}W_{i_2}$; (3) given two pairs $W_{i_1}W_{i_2}$ and $W_{j_1}W_{j_2}$ of individual constants of L^N with different indexes, let $W_{i_1}W_{i_2}$ precede (follow) $W_{j_1}W_{j_2}$ if the index of $W_{i_1}W_{i_2}$ is smaller than (is larger than) that of $W_{j_1}W_{j_2}$; (4) given two pairs $W_{i_1}W_{i_2}$ and $W_{j_1}W_{j_2}$ of individual constants of L^N with identical indexes, let $W_{i_1}W_{i_2}$ precede (follow) $W_{j_1}W_{j_2}$ if the index of W_{i_1} is smaller than (is larger than) that of W_{j_1}. The scheme is easily extended to suit sequences of three or more individual constants of L^N.

[5] The probability in question answers the one which would be allotted in Chapter 2 to the subset $\hat{w}\hat{x}$(w outlives x) of the so-called Cartesian product of an N-membered probability set PS by itself.

[6] The various instances of an open sentence of L^∞ may be true in L^∞ and yet be false in some (upon occasion, all) of the sublanguages of L^∞ of which they are sentences; hence my multiplying $\mathbf{w}^N(W_i)$ by $\mathbf{Ps}^\infty$ (W_i outlives Peter) rather than $\mathbf{Ps}^N$(W_i outlives Peter). See Note 15 of this chapter for further details on the matter. $\mathbf{Ps}^\infty$ (w outlives Peter), as reckoned here, may vary with the alphabetical order in which the individual constants of L^∞ are arranged in D1.1(g).

[7] The probability in question answers the one which would be allotted in Chapter 2 to the subset $\hat{w}\hat{x}$(w outlives x) of the so-called Cartesian product of a denumerably infinite and serially ordered probability set PS by itself; it may vary, of course, with the alphabetical order in which the individual constants of L^∞ are arranged in D1.1(g).

[8] I write '$\mathbf{Ps}^\infty$ (P, Q)' when I wish to emphasize that P and Q are sentences of L^∞, '$\mathbf{Ps}^N(P, Q)$' when I wish to emphasize that P and Q are sentences of L^N; otherwise, I write '$\mathbf{Ps}(P, Q)$,' as I have just done in the text.

[9] To condense matters, I pass over in this proof and later ones a number of obvious, but lengthy, mathematical transformations.

[10] Some chronological references may be in order here. In "Two Probability Concepts" (1956) I proposed for the sentences of L^1, L^2, L^3, and so on, the probability allotment detailed in theorems T13.6(a) and T13.6(c1)–(c2), and in "On chances and estimated chances of being true" (1959) I extended the allotment to cover as in theorems T13.6(b) and T13.6(d1)–(d2) the sentences of L^∞. In "Statistical and Inductive Probabilities," a paper read in early 1961 at The Pennsylvania State University, Wesleyan University, and Georgetown University, I generalized my 1956 and 1959 allotment into the various allotments detailed in theorem T13.5. At the urging of Professors Richard Jeffrey and Stig Kanger,

I now generalize my 1961 allotments into the various allotments detailed in definitions D13.1–2. A probability allotment roughly similar to the T13.6 one will be found in K. Ajdukiewicz, "La notion de rationalité des méthodes d'inférence faillibles" (1959), and, to merely jot down names, in B. Bolzano, C. S. Peirce, B. Russell, and D. C. Williams. Allotments roughly similar to the T13.5 ones (and, up to a point, the D13.1–2 ones) will be found in Kemeny, Mirkil, Snell, and Thompson, *Finite Mathematical Structures*, 1959; the authors restrict themselves, however, to open sentences hailing from a finite language and containing free occurrences of at most one individual variable; they also allot closed sentences other probabilities besides 1 and 0, a matter which I shall take up in Section 16.

[11] A sum of terms is said to be weighted when a certain weight (here the weight in L^N of a certain sequence of individual constants of L^N) is attached to each one of its terms.

[12] As the names jotted down in Note 10 of this chapter testify.

[13] Some of the material of this section is borrowed from my paper "On chances and estimated chances of being true."

[14] For the reason given in Note 6 of this chapter, T15.1 would not hold for all open sentences of L^∞ if $\mathbf{w}^N(W_{i_1}W_{i_2}\cdots W_{i_n})$ were multiplied in D13.2(d) by $\mathbf{Ps}^N(P_i)$ rather than $\mathbf{Ps}^\infty(P_i)$.

[15] Let indeed P be a sentence of L in which no individual variable of L distinct from X is free. It then follows from Chapter 1 that if every instance of P in L is true in L, $\hat{X}P$ must designate $\vee$ in L.

[16] I owe to Professor Ullian the idea of proving T15.4 by mathematical induction on $\max(p, q)$. For a different—and considerably longer—proof of T15.4, see "On chances and estimated chances of being true," pp. 232–234, where T3.8, a theorem which serves in the paper in question as a lemma to T15.4, calls for a slight correction.

[17] Let indeed P and Q be two sentences of L in which no individual variable of L distinct from W is free. It then follows from Chapter 1 that (1) if every instance of $\sim (P \,\&\, Q)$ in L is true in L, the two sets respectively designated in L by $\hat{W}P$ and $\hat{W}Q$ do not overlap, and (2) the set designated in L by $\hat{W}(P \vee Q)$ is the union of the two sets respectively designated in L by $\hat{W}P$ and $\hat{W}Q$.

[18] The logical equivalence in L of P and $W = W \supset P$ is easily proved by means of D4.1, D4.5, and D2.9.

[19] I assume in both cases that '$\mathbf{Tv}(P_i)$' has been substituted for '$\mathbf{Ps}(P_i)$' in D13.2(c)–(d) and, as a result, that the open sentences of L retain the probabilities they were allotted in Section 13.

[20] See Kemeny, Mirkil, Snell, and Thompson, *loc. cit.*, pp. 60–68 and 112–114. The reference to Kemeny in the text is again an abbreviated reference to all four authors. Examples of one-place functors are 'The size of,' 'The temperature of,' and so on.

[21] See the remarks appended to D3.5.

[22] For further details, see the author's "On a recent allotment of probabilities to open and closed sentences."

[23] I assume in all three cases that the individual constants of L^N are all weighted alike.

[24] A similar difficulty is discussed in Russell, *loc. cit.*, p. 354.

[25] Russell, *loc. cit.*, p. 354, handles Case 1 somewhat differently; he does not consider Case 2.

[26] The allotment was recommended to me, in the rough form in which it appears on pages 87–88, by the late Professor Arthur Pap. For a somewhat similar suggestion, see J. H. Lenz, "The Frequency Theory of Probability."

[27] To simplify matters here, I take all non-deductive inferences to be inductive ones. Many writers use the label 'inductive inference' more discriminately.

[28] I assume throughout that at least one instance of P' in L^N is true in L^N and hence by T14.2(a)–(b) that $\mathbf{Ps}^N(P')$ is non-zero.

[29] The function **CRs** is foreshadowed in the Ajdukiewicz article mentioned in Note 10 of this chapter.

4 INDUCTIVE PROBABILITIES

In this closing chapter I turn to inductive probabilities. First, I give instructions (some borrowed from Rudolf Carnap, some novel) for allotting inductive probabilities to the sentences of L (Section 18); review three major variants of those instructions (Sections 19–20); and show that inductive probabilities qualify as estimates—made in the light of one sentence—of the truth-value of another (Section 21). Then, I discuss the role played by logical falsehoods and logical truths as so-called evidence sentences (Section 22); match the personal probabilities of Frank P. Ramsey, Bruno de Finetti, and Leonard J. Savage against inductive probabilities (Section 23); and take a last look at inductive inferences (Section 24).

18. INDUCTIVE PROBABILITIES ALLOTTED TO THE SENTENCES OF THE LANGUAGES L

Pairs of closed sentences of L have been allotted inductive probabilities by Carnap and others in the following fashion. A list of requirements (reminiscent, as it happens, of T15.7, T15.8, T15.12, T15.13, T15.10, and T15.11) is first drawn up, and a function $\mathbf{Pi}^N$ taking pairs of closed sentences of L^N as its arguments, taking real numbers as its values, and meeting the said requirements,[1] is presumed to be on hand for each sublanguage L^N of L^∞. $\mathbf{Pi}^\infty(P, Q)$, where P and Q are two closed sentences of L^∞, is then taken to be the limit (if any) for increasing i of $\mathbf{Pi}^{N+i}(P, Q)$, where L^N is the first sublanguage of L^∞ of which both P and Q are sentences.[2]

Further pairs of sentences of L have recently been allotted inductive probabilities in the following manner. First, when P is an open sentence and Q a closed sentence of L^N, $\mathbf{Pi}^N(P, Q)$ has been taken to be the sum

$$\sum_{i=1}^{N} (\mathbf{w}^N(W_i) \cdot \mathbf{Pi}^N(P_i, Q)), \tag{1}$$

where, for each i from 1 to N, W_i is the i-th individual constant of L^N and P_i is like P except for containing occurrences of W_i at all the places where P contains free occurrences of some individual variable of L^N. Next, when P is an open sentence and Q a closed sentence of L^∞, $\mathbf{Pi}^\infty(P, Q)$ has been taken to be the limit (if any) for increasing N of the sum

$$\sum_{i=1}^{N} (\mathbf{w}^N(W_i) \cdot \mathbf{Pi}^\infty(P_i, Q)), \tag{2}$$

where, for each N from 1 on and each i from 1 to N, W_i and P_i are as in (1).[3]

These suggestions may be generalized as follows:

D18.1. (a) *For each sublanguage L^N of L^∞, a function $\mathbf{Pi}^N$ taking pairs of closed sentences of L^N as its arguments, taking real numbers as its values, and meeting the six requirements which follow, is presumed to be on hand:*

(a1) $0 \leq \mathbf{Pi}^N(P, Q)$;
(a2) $\mathbf{Pi}^N(P, P) = 1$;

(a3) *If P and Q are logically equivalent in L^N,*[4] *then*

$$\mathbf{Pi}^N(P, R) = \mathbf{Pi}^N(Q, R);$$

(a4) *If Q and R are logically equivalent in L^N,*[5] *then*

$$\mathbf{Pi}^N(P, Q) = \mathbf{Pi}^N(P, R);$$

(a5) $\mathbf{Pi}^N(P \,\&\, Q, R) = \mathbf{Pi}^N(P, Q \,\&\, R) \cdot \mathbf{Pi}^N(Q, R);$

(a6) *If Q is not logically false in L^N,*[6] *then*

$$\mathbf{Pi}^N(\sim P, Q) = 1 - \mathbf{Pi}^N(P, Q).$$ [7]

(b) *Let P and Q be two closed sentences of L^∞ and let L^N be the first sublanguage of L^∞ of which both P and Q are sentences. Then*

$$\mathbf{Pi}^\infty\,(P, Q) = \underset{i \to \infty}{Limit}\ \mathbf{Pi}^{N+i}(P, Q)$$

if such a limit exists; otherwise $\mathbf{Pi}^\infty$ *(P, Q) has no value.*

(c) *Let P be an open sentence and Q be a closed sentence of L^N; let $X_1, X_2, \cdots$, and X_n be the n ($n \geq 1$) individual variables of L^N which are free in P; and for each i from 1 to N^n let $W_{i_1}W_{i_2} \cdots W_{i_n}$ be the i-th of the sequences made up of any n individual constants of L^N and P_i be like P except for containing occurrences of $W_{i_1}, W_{i_2}, \cdots$, and W_{i_n} at all the places where P contains free occurrences of $X_1, X_2, \cdots$, and X_n, respectively. Then*

$$\mathbf{Pi}^N(P, Q) = \sum_{i=1}^{N^n} (\mathbf{w}^N(W_{i_1}W_{i_2} \cdots W_{i_n}) \cdot \mathbf{Pi}^N(P_i, Q)).$$

(d) *Let P be an open sentence and Q be a closed sentence of L^∞; let $X_1, X_2, \cdots$, and X_n be the n ($n \geq 1$) individual variables of L^∞ which are free in P; and for each N from 1 on and each i from 1 to N^n let $W_{i_1}W_{i_2} \cdots W_{i_n}$ and P_i be as in* (c). *Then*

$$\mathbf{Pi}^\infty\,(P, Q) = \underset{N \to \infty}{Limit} \sum_{i=1}^{N^n} (\mathbf{w}^N(W_{i_1}W_{i_2} \cdots W_{i_n}) \cdot \mathbf{Pi}^\infty\,(P_i, Q))$$

if such a limit exists; otherwise $\mathbf{Pi}^\infty$ *(P, Q) has no value.*

Further clauses, covering the case where Q in $\mathbf{Pi}^N(P, Q)$ is an open sentence of L^N and the one where Q in $\mathbf{Pi}^\infty\,(P, Q)$ is an open sentence of L^∞, may be added to D18.1; they are of little interest, however. It should be noticed that when, for each N from 1 on, each sequence of n ($n \geq 2$) individual constants of L^N is allotted as its weight in L^N the product of the weights in L^N of the n constants in question, the sum shown in D18.1(c) reduces to (1) on page 98 and the limit shown in D18.1(d) reduces to the limit—for increasing N—of (2) on that page.[8]

The reader may illustrate for himself the various clauses of D18.1(a) by letting P throughout read 'Mary outlives Peter,' Q in clauses (a1), (a4), and (a6) read 'Mary and Peter were born the same year,' Q in clause (a3) read 'Mary outlives both Peter and John or outlives Peter but not John,' Q in clause (a5) read 'Mary outlives John,' R in clause (a3) read 'Mary and Peter were born the same year,' R in clause (a4) read 'Mary and Peter were not born on different years,' and R in clause (a5) read 'Mary, Peter, and John were born the same year.'

The following consequences of D18.1(a) pave the way for the two variants of D18.1(a) to be studied in Section 19.

T18.2. *Let P, Q, and R be closed sentences of L^N.*

(a) *If Q logically implies P in L^N, then* $\mathbf{Pi}^N(P, Q) = 1$;

(b) *If P is logically true in L^N, then* $\mathbf{Pi}^N(P, Q) = 1$;

(c) *If Q is logically false in L^N, then* $\mathbf{Pi}^N(P, Q) = 1$;

(d) *If R is not logically false in L^N, then*

$$\mathbf{Pi}^N(P, R) = \mathbf{Pi}^N(P \mathbin{\&} Q, R) + \mathbf{Pi}^N(P \mathbin{\&} \sim Q, R);$$

(e) *If R is not logically false in L^N and R logically implies $\sim (P \mathbin{\&} Q)$ in L^N, then*

$$\mathbf{Pi}^N(P \vee Q, R) = \mathbf{Pi}^N(P, R) + \mathbf{Pi}^N(Q, R).$$

Proof: (a) If Q logically implies P in L^N, then by D2.7, D2.9, D4.1, and D4.4–5 $P \mathbin{\&} Q$ is logically equivalent to Q in L^N, hence by D18.1(a3) and D18.1(a2) $\mathbf{Pi}^N(P \mathbin{\&} Q, Q) = 1$, and hence by D18.1(a5) and D18.1(a2) $\mathbf{Pi}^N(P, Q \mathbin{\&} Q) = 1$. But by D2.7, D2.9, D4.1, and D4.5 $Q \mathbin{\&} Q$ is logically equivalent to Q in L^N. Hence (a) by D18.1(a4).

(b) If P is logically true in L^N, then by D4.1–2 and D4.4 Q logically implies P in L^N. Hence (b) by (a).

(c) If Q is logically false in L^N, then by D4.1 and D4.3–4 Q logically implies P in L^N. Hence (c) by (a).

(d) By D18.1(a6) in case $P \mathbin{\&} R$ is not logically false in L^N and by (c) in case $P \mathbin{\&} R$ is logically false in L^N,

$$\mathbf{Pi}^N(R, P \,\&\, R) + \mathbf{Pi}^N(\sim R, P \,\&\, R) = \mathbf{Pi}^N(Q, P \,\&\, R) + \mathbf{Pi}^N(\sim Q, P \,\&\, R),$$

and hence by D18.1(a5)

$$\mathbf{Pi}^N(R \,\&\, P, R) + \mathbf{Pi}^N(\sim R \,\&\, P, R) = \mathbf{Pi}^N(Q \,\&\, P, R) + \mathbf{Pi}^N(\sim Q \,\&\, P, R).$$

But by D2.7, D2.9, D4.1, and D4.5 $Q \,\&\, P$ is logically equivalent in L^N to $P \,\&\, Q$, $\sim Q \,\&\, P$ to $P \,\&\, \sim Q$, $R \,\&\, P$ to $P \,\&\, R$, $\sim R \,\&\, P$ to $P \,\&\, \sim R$, and $R \,\&\, R$ to R. Hence by D18.1(a3), D18.1(a5), and D18.1(a2)

$$\mathbf{Pi}^N(P, R) + (\mathbf{Pi}^N(P, \sim R \,\&\, R) \cdot \mathbf{Pi}^N(\sim R, R)) = \mathbf{Pi}^N(P \,\&\, Q, R) + \mathbf{Pi}^N(P \,\&\, \sim Q, R).$$

But if R is not logically false in L^N, then by D18.1(a2) and D18.1(a6) $\mathbf{Pi}^N(\sim R, R) = 0$. Hence (d).[9]

(e) If R is not logically false in L^N, then by (d)

$$\mathbf{Pi}^N(P \vee Q, R) = \mathbf{Pi}^N((P \vee Q) \,\&\, Q, R) + \mathbf{Pi}^N((P \vee Q) \,\&\, \sim Q, R).$$

But by D2.7–9, D4.1, and D4.5 $(P \vee Q) \,\&\, Q$ is logically equivalent in L^N to Q and $(P \vee Q) \,\&\, \sim Q$ to $P \,\&\, \sim Q$. Hence if R is not logically false in L^N, then by D18.1(a3) and (d)

$$\mathbf{Pi}^N(P \vee Q, R) = \mathbf{Pi}^N(P, R) + \mathbf{Pi}^N(Q, R) - \mathbf{Pi}^N(P \,\&\, Q, R).$$

But if R is not logically false in L^N and R logically implies $\sim (P \,\&\, Q)$ in L^N, then by (a) and D18.1(a6) $\mathbf{Pi}^N(P \,\&\, Q, R) = 0$. Hence (e).[10]

T18.3. *Let P and Q be closed sentences of L^N and W be an individual constant of L^N.*

(a) $0 \leq \mathbf{Pi}^N(P, W = W)$;

(b) *If P is logically true in L^N, then* $\mathbf{Pi}^N(P, W = W) = 1$;

(c) *If P and Q are logically equivalent in L^N, then*

$$\mathbf{Pi}^N(P, W = W) = \mathbf{Pi}^N(Q, W = W);$$

(d) *If $P \,\&\, Q$ is logically false in L^N, then*

$$\mathbf{Pi}^N(P \vee Q, W = W) = \mathbf{Pi}^N(P, W = W) + \mathbf{Pi}^N(Q, W = W);$$

(e) *If* $\mathbf{Pi}^N(Q, W = W) \neq 0$, *then*

$$\mathbf{Pi}^N(P, Q) = \mathbf{Pi}^N(P \,\&\, Q, W = W)/\mathbf{Pi}^N(Q, W = W).$$

Proof: (a) by D18.1(a1).

(b) by T18.2(b).

(c) by D18.1(a3).

(d) If $P \,\&\, Q$ is logically false in L^N, then by D4.1 and D4.3–4 $W = W$ logically implies $\sim (P \,\&\, Q)$ in L^N. But by D3.2–3, D4.1, and D4.3 $W = W$ is not logically false in L^N.[11] Hence (d) by T18.2(d).

(e) By D2.7, D2.9, D4.1, and D4.5 $Q \,\&\, W = W$ is logically equivalent to Q in L^N. Hence by D18.1(a5) and D18.1(a4)

$$\mathbf{Pi}^N(P, Q) \cdot \mathbf{Pi}^N(Q, W = W) = \mathbf{Pi}^N(P \,\&\, Q, W = W).$$

Hence (e).

T18.4. *Let P, Q, and R be closed sentences of L^N.*

(a) $\mathbf{Pi}^N(P \,\&\, Q, R) = \mathbf{Pi}^N(Q \,\&\, P, R)$;

(b) *If* $\mathbf{Pi}^N(Q, S) = \mathbf{Pi}^N(R, S)$ *for every closed sentence S of L^N, then*

$$\mathbf{Pi}^N(P, Q) = \mathbf{Pi}^N(P, R);$$

(c) *If* $\mathbf{Pi}^N(R, Q) \neq 1$ *for at least one closed sentence R of L^N, then*

$$\mathbf{Pi}^N(\sim P, Q) = 1 - \mathbf{Pi}^N(P, Q);$$

(d) *P is logically true in L^N if and only if* $\mathbf{Pi}^N(P, Q) = 1$ *for every closed sentence Q of L^N.*

Proof: (a) By D2.7, D2.9, D4.1, and D4.5 $P \,\&\, Q$ is logically equivalent to $Q \,\&\, P$ in L^N. Hence (a) by D18.1(a3).

(b) 1. If $\mathbf{Pi}^N(Q, S) = \mathbf{Pi}^N(R, S)$ for every closed sentence S of L^N, then

$$\mathbf{Pi}^N(Q, Q) = \mathbf{Pi}^N(R, Q),$$
$$\mathbf{Pi}^N(Q, R) = \mathbf{Pi}^N(R, R),$$

and hence by D18.1(a2)

$$\mathbf{Pi}^N(R, Q) = \mathbf{Pi}^N(Q, R).$$

But by D2.7, D2.9, D4.1, and D4.5 $Q \,\&\, R$ is logically equivalent to $R \,\&\, Q$ in L^N, and hence by D18.1(a4)

$$\mathbf{Pi}^N(P, Q \,\&\, R) = \mathbf{Pi}^N(P, R \,\&\, Q).$$

Hence if $\mathbf{Pi}^N(Q, S) = \mathbf{Pi}^N(R, S)$ for every closed sentence S of L^N, then by D18.1(a5)

$$\mathbf{Pi}^N(P \,\&\, Q, R) = \mathbf{Pi}^N(P \,\&\, R, Q),$$

hence by (a)

$$\mathbf{Pi}^N(Q \,\&\, P, R) = \mathbf{Pi}^N(R \,\&\, P, Q),$$

and hence by D18.1(a5)

$$\mathbf{Pi}^N(Q, P \,\&\, R) \cdot \mathbf{Pi}^N(P, R) = \mathbf{Pi}^N(R, P \,\&\, Q) \cdot \mathbf{Pi}^N(P, Q).$$

2. If $\mathbf{Pi}^N(Q, S) = \mathbf{Pi}^N(R, S)$ for every closed sentence S of L^N, then

$$\mathbf{Pi}^N(Q, P \,\&\, R) = \mathbf{Pi}^N(R, P \,\&\, R)$$

and

$$\mathbf{Pi}^N(Q, P \,\&\, Q) = \mathbf{Pi}^N(R, P \,\&\, Q).$$

But by D2.7, D4.1, and D4.4 $P \,\&\, R$ logically implies R in L^N, $P \,\&\, Q$ logically implies Q in L^N, and hence by T18.2(a)

$$\mathbf{Pi}^N(R, P \,\&\, R) = \mathbf{Pi}^N(Q, P \,\&\, Q),$$

where $\mathbf{Pi}^N(R, P \,\&\, R) \neq 0$. Hence, if $\mathbf{Pi}^N(Q, S) = \mathbf{Pi}^N(R, S)$ for every closed sentence S of L^N, then

$$\mathbf{Pi}^N(Q, P \,\&\, R) = \mathbf{Pi}^N(R, P \,\&\, Q),$$

where $\mathbf{Pi}^N(Q, P \,\&\, R) \neq 0$.

3. (b) by 1 and 2.[12]

(c) If $\mathbf{Pi}^N(R, Q) \neq 1$ for at least one closed sentence R of L^N, then by T18.2(c) Q is not logically false in L^N. Hence (c) by D18.1(a6).

(d) 1. If P is logically true in L^N, then by T18.2(b) $\mathbf{Pi}^N(P, Q) = 1$ for every closed sentence Q of L^N.

2. If $\mathbf{Pi}^N(P, Q) = 1$ for every closed sentence Q of L^N, then

$$\mathbf{Pi}^N(P, \sim P) = 1.$$

But by D4.1 and D4.5 P is logically equivalent to $\sim\sim P$ in L^N, and hence by D18.1(a3)

$$\mathbf{Pi}^N(P, \sim P) = \mathbf{Pi}^N(\sim\sim P, \sim P).$$

Hence, if $\mathbf{Pi}^N(P, Q) = 1$ for every closed sentence Q of L^N, then

$$\mathbf{Pi}^N(\sim\sim P, \sim P) = 1,$$

hence by D18.1(a6) and D18.1(a2) $\sim P$ is logically false in L^N, and hence by D4.1–3 P is logically true in L^N.

3. (d) by 1 and 2.

I should point out, before ending this section, that if a closed sentence P of L^∞ is logically true in L^∞, then P is logically true as well in every sublanguage of L^∞ from a given L^N on.[13] D18.1(a)–(b) thus yield the following consequences, where $\mathbf{Pi}^\infty(P, Q)$ is presumed to have a value for every pair of sentences P and Q concerned:

T18.5. *Let P, Q, and R be closed sentences of* L^{∞}.

(a) $0 \leq \mathbf{Pi}^{\infty}(P, Q)$;

(b) $\mathbf{Pi}^{\infty}(P, P) = 1$;

(c) *If P and Q are logically equivalent in* L^{∞}, *then*

$$\mathbf{Pi}^{\infty}(P, R) = \mathbf{Pi}^{\infty}(Q, R);$$

(d) *If Q and R are logically equivalent in* L^{∞}, *then*

$$\mathbf{Pi}^{\infty}(P, Q) = \mathbf{Pi}^{\infty}(P, R);$$

(e) $\mathbf{Pi}^{\infty}(P \,\&\, Q, R) = \mathbf{Pi}^{\infty}(P, Q \,\&\, R) \cdot \mathbf{Pi}^{\infty}(Q, R)$;

(f) *If Q is not logically false in any sublanguage of* L^{∞} *from a given* L^N *on, then*

$$\mathbf{Pi}^{\infty}(\sim P, Q) = 1 - \mathbf{Pi}^{\infty}(P, Q).$$

19. VARIANTS OF THE STANDARD ALLOTMENTS (I)

Numerous variants of D18.1(a) will be found in the literature, some equivalent to D18.1(a), some not. I study two of them in this section and a third one in the next.

The inductive probabilities I dealt with in D18.1(a) went to pairs of closed sentences of L^N and hence were of a conditional sort. Absolute inductive probabilities may also be allotted to single closed sentences of L^N, and $\mathbf{Pi}^N(P, Q)$ may then be reckoned by a procedure reminiscent of D13.3. Details are as follows, where D19.1(a)–(d) are reminiscent of T15.1, T15.2, T15.4, and T15.6.

D19.1. *For each sublanguage* L^N *of* L^{∞}, *a function* $\mathbf{Pi}_0^N$ *taking closed sentences of* L^N *as its arguments, taking real numbers as its values, and meeting the requirements which follow, is presumed to be on hand:*

(a) $0 \leq \mathbf{Pi}_0^N(P)$;

(b) *If P is logically true in* L^N, *then* $\mathbf{Pi}_0^N(P) = 1$;

(c) *If P and Q are logically equivalent in* L^N, *then*

$$\mathbf{Pi}_0^N(P) = \mathbf{Pi}_0^N(Q);$$

(d) *If* $P \,\&\, Q$ *is logically false in* L^N, *then*

$$\mathbf{Pi}_0^N(P \vee Q) = \mathbf{Pi}_0^N(P) + \mathbf{Pi}_0^N(Q).$$[14]

D19.2. (a) *Let P and Q be two closed sentences of* L^N.

(a1) *If* $\mathbf{Pi}_0^N(Q) \neq 0$, *then* $\mathbf{Pi}^N(P, Q) = \mathbf{Pi}_0^N(P \,\&\, Q)/\mathbf{Pi}_0^N(Q)$;

(a2) *If* $\mathbf{Pi}_0^N(Q) = 0$, *then* $\mathbf{Pi}^N(P, Q) = 1$.

(b)–(d) *Like D18.1(b)–(d).*

It is then possible to prove the following theorem, according to which $\mathbf{Pi}^N$ as defined in D19.2(a) satisfies each one of D18.1(a1)–(a5) plus a weakened version of D18.1(a6), and has for a pair of closed sentences P and $W = W$ of L^N the same value that $\mathbf{Pi}_0^N$ has for P.

T19.3. *Let* $\mathbf{Pi}^N$ *be as in D19.2(a); let* P, Q, *and* R *be closed sentences of* L^N; *and let* W *be an individual constant of* L^N.

(a) $0 \leq \mathbf{Pi}^N(P, Q)$;

(b) $\mathbf{Pi}^N(P, P) = 1$;

(c) *If* P *and* Q *are logically equivalent in* L^N, *then*

$$\mathbf{Pi}^N(P, R) = \mathbf{Pi}^N(Q, R);$$

(d) *If* Q *and* R *are logically equivalent in* L^N, *then*

$$\mathbf{Pi}^N(P, Q) = \mathbf{Pi}^N(P, R);$$

(e) $\mathbf{Pi}^N(P \,\&\, Q, R) = \mathbf{Pi}^N(P, Q \,\&\, R) \cdot \mathbf{Pi}^N(Q, R)$;

(f) *If* $\mathbf{Pi}^N(Q, W = W) \neq 0$, *then* $\mathbf{Pi}^N(\sim P, Q) = 1 - \mathbf{Pi}^N(P, Q)$;

(g) $\mathbf{Pi}^N(P, W = W) = \mathbf{Pi}_0^N(P)$.

Proof: (a)–(f) Proofs similar to the proofs of T15.7–8 and T15.10–13 (with the extra aid of (g) in the case of (f)).

(g) By D19.1(b) and D4.1–2 $\mathbf{Pi}_0^N(W = W) = 1$. But by D2.7, D2.9, D4.1, and D4.5 $P \,\&\, W = W$ is logically equivalent to P in L^N, and hence by D19.1(c) $\mathbf{Pi}_0^N(P \,\&\, W = W) = \mathbf{Pi}_0^N(P)$. Hence (g) by D19.2(a1).

The function $\mathbf{Pi}^N$ described in D18.1(a) does not exactly match the one defined in D19.2(a), since (1) by means of D19.2(a) requirement D18.1(a6) is provable only under the form T19.3(f), and (2) by means of D18.1(a) requirement D19.2(a2) is provable only under the form T18.2(c). The two functions come to match exactly, however, once the following requirements are added to D18.1(a) and D19.1, respectively:

D18.1. (a7) *If* $\mathbf{Pi}^N(P, Q) = 1$, *then* Q *logically implies* P *in* L^N,

and

D19.1. (e) *If* $\mathbf{Pi}_0^N(P) = 1$, *then* P *is logically true in* L^N.

T19.4. *Let* $\mathbf{Pi}^N$ *meet requirements D18.1(a1)–(a7); let* P *and* Q *be* *closed sentences of* L^N*; and let* W *be an individual constant of* L^N*. If* $\mathbf{Pi}^N(Q, W = W) = 0$*, then* $\mathbf{Pi}^N(P, Q) = 1$.

Proof: By D3.2–3, D4.1, and D4.3 $W = W$ is not logically false in L^N. Hence if $\mathbf{Pi}^N(Q, W = W) = 0$, then by D18.1(a6)

$$\mathbf{Pi}^N(\sim Q, W = W) = 1,$$

hence by D18.1(a7) $W = W$ logically implies $\sim Q$ in L^N, hence by D4.1 and D4.3–4 Q is logically false in L^N, and hence by T18.2(c)

$$\mathbf{Pi}^N(P, Q) = 1.$$

T19.5. *Let* $\mathbf{Pi}_0^N$ *meet requirements D19.1(a)–(e); let* $\mathbf{Pi}^N$ *be as in D19.2(a); and let* P *and* Q *be closed sentences of* L^N*. If* Q *is not logically false in* L^N*, then* $\mathbf{Pi}^N(\sim P, Q) = 1 - \mathbf{Pi}^N(P, Q)$.[15]

Proof: By D2.7, D4.1, and D4.3 $Q \mathbin{\&} \sim Q$ is logically false in L^N; hence by D19.1(d)

$$\mathbf{Pi}_0^N(Q \vee \sim Q) = \mathbf{Pi}_0^N(Q) + \mathbf{Pi}_0^N(\sim Q).$$

But by D2.8 and D4.1–2 $Q \vee \sim Q$ is logically true in L^N, and hence by D19.1(b) $\mathbf{Pi}_0^N(Q \vee \sim Q) = 1$. Hence if $\mathbf{Pi}_0^N(Q) = 0$, then $\mathbf{Pi}_0^N(\sim Q) = 1$, hence by D19.1(e) $\sim Q$ is logically true in L^N, and hence by D4.2–3 Q is logically false in L^N. But if $\mathbf{Pi}_0^N(Q, W = W) = 0$, where W is an individual constant of L^N, then by T19.3(g) $\mathbf{Pi}_0^N(Q) = 0$. Hence T19.5 by T19.3(f).

The two functions also come to match exactly when (1) requirement D18.1(a6) is weakened to read like T19.3(f), and (2) the following requirement is added to D18.1(a):

D18.1. (a8) *If* $\mathbf{Pi}^N(Q, W = W) = 0$*, then* $\mathbf{Pi}^N(P, Q) = 1$*, where* W *is an individual constant of* L^N.

Altering somewhat Carnap's terminology on this score, I shall call the various functions $\mathbf{Pi}^N$ which meet requirements D18.1(a1)–(a7) strongly regular, those which meet only requirements D18.1(a1)–(a6) regular, and those which meet only requirements T19.3(a)–(f) weakly regular.[16] It follows from the above that (1) if $\mathbf{Pi}_0^N$ meets requirements D19.1(a)–(e) and $\mathbf{Pi}^N$ is defined as in D19.2(a), then $\mathbf{Pi}^N$ is strongly regular, and (2) if $\mathbf{Pi}_0^N$ meets only requirements D19.1(a)–(d) and $\mathbf{Pi}^N$ is defined again as in D19.2(a), then $\mathbf{Pi}^N$ is weakly regular. The two results will prove of interest in Section 20.

Now for another variant of D18.1(a), a variant equivalent this time to D18.1(a). Let a sentence P of L^N be said to be a quantifier-and-identity-free sentence of L^N if P is of the form $G(W_1, W_2, \cdots, W_n)$, where G is an n-place ($n \geq 1$) predicate of L^N and $W_1, W_2, \cdots$, and W_n are n individual signs of L^N, or of the form $\sim Q$, where Q is a quantifier-and-identity-free sentence of L^N, or of the form $Q \supset R$, where Q and R are quantifier-and-identity-free sentences of L^N. It can be shown that to every closed sentence of L^N there corresponds a closed quantifier-and-identity-free sentence of L^N which is logically equivalent to it in L^N and hence that the function $\mathbf{Pi}^N$ described in D18.1(a) may, to all intents and purposes, be restricted to take pairs of closed quantifier-and-identity-free sentences of L^N as its arguments.[17] It can further be shown that when $\mathbf{Pi}^N$ is restricted to take pairs of closed quantifier-and-identity-free sentences of L^N as its arguments, $\mathbf{Pi}^N$ meets requirements D18.1(a1)–(a6) if and only if $\mathbf{Pi}^N$ meets requirements D18.1(a1), D18.1(a2), T18.4(a), T18.4(b), D18.1(a5), T18.4(c), and T18.4(d). It can finally be shown that when $\mathbf{Pi}^N$ is restricted to take pairs of closed quantifier-and-identity-free sentences of L^N as its arguments and D4.2 is modified to read as follows:

D4.2′. *A closed quantifier-and-identity-free sentence P of L^N is logically true in L^N if $\mathbf{Pi}^N(P, Q) = 1$ for every closed quantifier-and-identity-free sentence Q of L^N,*

then (1) $\mathbf{Pi}^N$ meets requirements D18.1(a1)–(a6) if and only if $\mathbf{Pi}^N$ meets requirements D18.1(a1), D18.1(a2), T18.4(a), T18.4(b), D18.1(a5), and T18.4(c), and (2) a closed quantifier-and-identity-free sentence P of L^N is logically true in L^N by virtue of D4.1–2 if and only if P is logically true in L^N by virtue of D4.2′. It follows from (2), by the way, that the notion of logical truth in L^N and the derivative notions of logical falsehood, logical implication, and logical equivalence in L^N are definable by means of the inductive probability function $\mathbf{Pi}^N$ when the said function is regular.[18]

20. VARIANTS OF THE STANDARD ALLOTMENTS (II)

Another way yet of allotting inductive probabilities to the closed sentences of L^N has been suggested by Carnap.[19] It calls for four preliminary definitions, D20.1–4. In the first two I indicate what is understood by an elementary sentence of L^N and a state-description of L^N; in the third I lay down the conditions under which a closed sentence of L^N is to hold in a state-description of L^N; in the fourth I introduce a function, the so-called holding-value function, whose value $\mathbf{Hv}(P, SD)$ for a closed sentence P of L^N and a state-description SD of L^N is to be 1 if P holds in SD, 0 if it does not.

D20.1. *A sentence P of L^N is an elementary sentence of L^N if P is of the form $G(W_1, W_2, \cdots, W_n)$, where G is an n-place ($n \geq 1$) predicate of L^N and $W_1, W_2, \cdots$, and W_n are n individual signs of L^N.*

Let L^4, for example, have only one predicate, a one-place predicate which I abbreviate as 'G,' and let 'a,' 'b,' 'c,' and 'd' be short for the four individual constants of L^4. The closed elementary sentences of L^4 will then be 'G(a),' 'G(b),' 'G(c),' and 'G(d).' I shall use 'α_N' to designate the number of closed elementary sentences of L^N.

D20.2. *A sentence P of L^N is a state-description of L^N if (a) P is $(\cdots(Q_1 \& Q_2) \& \cdots) \& Q_{\alpha_N}$, where $Q_1, Q_2, \cdots$, and Q_{α_N} are in some arbitrary order the α_N closed elementary sentences of L^N, or if (b) P is the result of inserting '$\sim$' before one or more of the sentences $Q_1, Q_2, \cdots$, and Q_{α_N} in $(\cdots(Q_1 \& Q_2) \& \cdots) \& Q_{\alpha_N}$.*[20]

Let L^4, for example, be as above. The state-descriptions of L^4 will then be as follows:

SD_1: ((G(a) & G(b)) & G(c)) & G(d)
SD_2: ((~ G(a) & G(b)) & G(c)) & G(d)
SD_3: ((G(a) & ~ G(b)) & G(c)) & G(d)
SD_4: ((~ G(a) & ~ G(b)) & G(c)) & G(d)
SD_5: ((G(a) & G(b)) & ~ G(c)) & G(d)
SD_6: ((~ G(a) & G(b)) & ~ G(c)) & G(d)
SD_7: ((G(a) & ~ G(b)) & ~ G(c)) & G(d)
SD_8: ((~ G(a) & ~ G(b)) & ~ G(c)) & G(d)
SD_9: ((G(a) & G(b)) & G(c)) & ~ G(d)
SD_{10}: ((~ G(a) & G(b)) & G(c)) & ~ G(d)
SD_{11}: ((G(a) & ~ G(b)) & G(c)) & ~ G(d)
SD_{12}: ((~ G(a) & ~ G(b)) & G(c)) & ~ G(d)
SD_{13}: ((G(a) & G(b)) & ~ G(c)) & ~ G(d)
SD_{14}: ((~ G(a) & G(b)) & ~ G(c)) & ~ G(d)

SD_{15}: ((G(a) & ~ G(b)) & ~ G(c)) & ~ G(d)
SD_{16}: ((~ G(a) & ~ G(b)) & ~ G(c)) & ~ G(d).

The number of state-descriptions of L^N is easily shown to be 2^{α_N}, where α^N is the number of closed elementary sentences of L^N. I shall use 'β_N' as an abbreviation for '2^{α_N}'; I shall also let '*SD*' range over the state-descriptions of L^N.

D20.3. *Let SD be a state-description of* L^N.

(a) *If* $G(W_1, W_2, \cdots, W_n)$, *where G is an n-place* ($n \geq 1$) *predicate of* L^N *and* $W_1, W_2, \cdots$, *and* W_n *are n individual constants of* L^N, *occurs in SD and is not preceded in SD by '~,' then* $G(W_1, W_2, \cdots, W_n)$ *holds in SD;*

(b) $W = W$, *where W is an individual constant of* L^N, *holds in SD;*

(c) *If P is a closed sentence of* L^N *and P does not hold in SD, then* ~ *P holds in SD;*

(d) *If P and Q are closed sentences of* L^N *and if P does not hold in SD and/or Q holds in SD, then* $P \supset Q$ *holds in SD;*

(e) *If no individual variable of* L^N *distinct from W is free in P and every instance of P in* L^N *holds in SD, then* $(\forall W)P$ *holds in SD;*

(f) *No closed sentence of* L^N *holds in SD unless its doing so follows from (a)–(e).*

It can easily be shown of any closed sentence *P* of L^N that (1) if *P* is logically false in L^N, then *P* does not hold in any state-description of L^N, and (2) if *P* is not logically false in L^N, then *P* holds in one or more state-descriptions of L^N. Let L^4 and SD_1–SD_{16}, for example, be as before. Then 'G(a) & ~ G(a)' does not hold in any one of SD_1–SD_{16}; '$(\forall w)$G(w)' holds in SD_1 (and SD_1 only); and '~ (G(a) $\supset$ ~ G(b))' holds in SD_1, SD_5, SD_9, and SD_{13} (and none other).

D20.4. *Let P be a closed sentence and SD a state-description of* L^N.

(a) *If P holds in SD, then* **Hv**(*P*, *SD*) *equals 1;*

(b) *If P does not hold in SD, then* **Hv**(*P*, *SD*) *equals 0.*

As weights were allotted in Chapter 3 to the individual constants of L^N, so weights may be allotted here to the state-descriptions of L^N. $\mathbf{Pi}_0^N(P)$, where *P* is a closed sentence of L^N, may then be taken to be 0 when *P* is logically false in L^N and to be the combined weights of the state-descriptions of L^N in which *P* holds when *P* is not logically false in L^N. $\mathbf{Pi}^N(P, Q)$, where P and Q are closed sentences of L^N, may finally be reckoned as in D19.2(a). Details are as follows:

D20.5. *For each N from 1 on, clauses of the following kind:*

(1) SD_1 *has* $\mathbf{w}^N(SD_1)$ *as its weight in* L^N,
(2) SD_2 *has* $\mathbf{w}^N(SD_2)$ *as its weight in* L^N,
.
.
(β_N) SD_{β_N} *has* $\mathbf{w}^N(SD_{\beta_N})$ *as its weight in* L^N,

are presumed to be on hand, where

(a) $SD_1, SD_2, \cdots$, *and* SD_{β_N} *are in some arbitrary order the various state-descriptions of the sublanguage* L^N *of* L^∞

and

(b) $\mathbf{w}^N(SD_1), \mathbf{w}^N(SD_2), \cdots$, *and* $\mathbf{w}^N(SD_{\beta_N})$ *are real numbers such that:*

(b1) $0 < \mathbf{w}^N(SD_i)$ *for each i from 1 to* SD_{β_N}

and

(b2) $\sum_{i=1}^{\beta_N} \mathbf{w}^N(SD_i) = 1$.

D20.6. *Let P be a closed sentence of* L^N.
(a) *If P is logically false in* L^N, *then* $\mathbf{Pi}_0^N(P) = 0$;
(b) *If P is not logically false in* L^N, *then*

$$\mathbf{Pi}_0^N(P) = \sum_{i=1}^{\beta_N} (\mathbf{w}^N(SD_i) \cdot \mathbf{Hv}(P, SD_i)).$$

D20.7. *Let P and Q be two closed sentences of* L^N.
(a) *If* $\mathbf{Pi}_0^N(Q) \neq 0$, *then* $\mathbf{Pi}^N(P, Q) = \mathbf{Pi}_0^N(P \,\&\, Q)/\mathbf{Pi}_0^N(Q)$;
(b) *If* $\mathbf{Pi}_0^N(Q) = 0$, *then* $\mathbf{Pi}^N(P, Q) = 1$.

It should be noted that (1) the state-descriptions of L^N correspond to Kemeny's logical possibilities, (2) the state-descriptions of L^N in which a closed sentence P of L^N holds correspond to those of Kemeny's logical possibilities which are not precluded by P, and (3) the inductive probability allotted in D20.6 to a closed sentence P of L^N corresponds to the probability allotted by Kemeny to P. Hence my previous claim that the probabilities allotted by Kemeny to closed sentences have an inductive flavor.[21]

In analogy with D13.1(b1), line (b1) in D20.5 is sometimes modified to read:

(b1′) *$0 \leq \mathbf{w}^N(SD_i)$ for each i from 1 to β_N.*

With (b1) as originally stated, the function $\mathbf{Pi}_0^N$ defined in D20.6 meets all five of requirements D19.1(a)–(e), and the weights allotted in D20.5 to SD_1, SD_2, $\cdots$, and SD_{β_N} may, as a result, be called strongly regular. With (b1) modified to read like (b1′), the function $\mathbf{Pi}_0^N$ in question meets only requirements D19.1(a)–(d), and the weights allotted in D20.5 to SD_1, SD_2, $\cdots$, and SD_{β_N} may, as a result, be called weakly regular.[22]

It can be shown, by about the same argument as on page 35, that, for any N from 1 on, weights meeting restrictions D20.5(b1) (or D20.5(b1′)) and D20.5(b2) can be allotted in L^N to the state-descriptions of L^N in $2^{\aleph_0}$ different ways. Among the allotments in question a number stand out, the so-called symmetrical allotments, under which

$$\mathbf{Pi}_0^N(P) = \mathbf{Pi}_0^N(P'),$$

where P is any closed sentence of L^N and P' is like P except for containing occurrences of an individual constant W' of L^N at those and only those places where P contains occurrences of an individual constant W of L^N. Carnap, who puts all the individual constants of L^N on a par, endorses only symmetrical allotments.[23] A few other writers are more catholic-minded. I, for one, recently suggested that when P is of the form $G(W)$, $\mathbf{Pi}_0^N(P) - \mathbf{Pi}_0^N(P')$ be smaller than, equal to, or larger than 0 according as the weight of W in L^N is smaller than, equal to, or larger than the weight of W' in L^N, and hence that the weights allotted in D20.5 to the state-descriptions of L^N vary up to a point with the weights allotted in D13.1 to the individual constants of L^N. The suggestion is too tentative, however, to be taken up here.[24]

Among Carnap's symmetrical allotments one stands out, the equiprobable or relative frequency allotment, under which

(1) each state-description of L^N has as its weight in L^N the ratio $1/\beta_N$ and hence

(2) each closed sentence P of L^N has as its inductive probability in

L^N the proportion, say m/β_N ($m \geq 0$), of state-descriptions of L^N in which P holds.[25]

Something like the equiprobable or relative frequency allotment was favored, it would seem, by Charles Sanders Peirce, John Maynard Keynes, and Ludwig Wittgenstein.[26]

21. INDUCTIVE PROBABILITIES AS ESTIMATES OF TRUTH-VALUES

$\mathbf{Pi}^N(P, Q)$ and, by extension, $\mathbf{Pi}^\infty(P, Q)$ are susceptible of various interpretations. I submit here that:

(1) when P and Q are two closed sentences of L^N, $\mathbf{Pi}^N(P, Q)$ qualifies as an estimate—made in the light of the information conveyed by Q or, for short, in the light of Q—of the truth-value $\mathbf{Tv}^N(P)$ of P in L^N, and

(2) when P is an open sentence and Q a closed sentence of L^N, $\mathbf{Pi}^N(P, Q)$ qualifies as an estimate—made in the light of Q—of the statistical probability $\mathbf{Ps}^N(P)$ of P in L^N.[27]

Since $\mathbf{Tv}^N(P)$ is equal by T14.1(a) to the statistical probability of P in L^N, I thus submit that $\mathbf{Pi}^N(P, Q)$ qualifies as an estimate—made in the light of Q—of the statistical probability of P in L^N. Since, on the other hand, $\mathbf{Ps}^N(P)$ is equal by T14.1(c) to a weighted sum of the truth-values in L^N of the instances of P in L^N and hence to the truth-value, in a generalized sense of the word, of P in L^N, I also submit that $\mathbf{Pi}^N(P, Q)$ qualifies as an estimate—made in the light of Q—of the truth-value of P in L^N.

The interpretation, in either wording, may be of interest: theoretical interest, first, in that it effects some kind of a rapprochement between statistical probabilities and inductive ones; practical interest, second, in that it provides gainful employment for inductive probabilities. It is also in keeping with my contention in the Preface that inductive probabilities betoken some uncertainty on our part as to the exact truth-value of the sentences to which they are allotted.[28]

The spectrum of functions by means of which the truth-value in L^N of a closed sentence P of L^N can be estimated, clearly runs from the function $\mathbf{Pi}^N$ defined in theorem T21.1 to the one defined in theorem T21.2. By means of the former P is automatically estimated to be false in L^N unless Q—the closed sentence of L^N in the light of which the truth-value of P in L^N is estimated—logically implies P in L^N. By means of the latter, P is automatically estimated to be true in L^N unless Q—the closed sentence of L^N in the light of which the truth-value of P in L^N is estimated—is not logically false in L^N and logically implies $\sim P$ in L^N. I first show, in behalf of (1), that the two functions in question, diametrically opposed though they may be, meet requirements D18.1(a1)–(a5), and make bold to claim, on the strength of the result, that the value for a pair of closed sentences P and Q of L^N of any function which likewise meets those five requirements must rate "pass or better" as an estimate—made in the light of Q—of the truth-value of P in L^N (T21.1–2). I next show that the two functions in question, extreme as they are, fail to meet requirement D18.1(a6), and make bold to claim, on the strength of the result, that the value for a pair of closed sentences P and Q of L^N of any function which meets all six of requirements D18.1(a1)–(a6) must rate "fair or better" as an estimate—made in the light of Q—of the truth-value of P in L^N (T21.3–4). Further requirements should clearly be added to D18.1(a1)–(a6) to weed out any function whose value for a pair of closed sentences P and Q of L^N, though rating "fair or better," rates no better than "fair" as an estimate—made in the light of Q—of the truth-value of P in L^N. The problem, however, need not concern us right now.

T21.1. *Let* $\mathbf{Pi}^N$ *be such that:*

(a) *If Q is logically false in L^N,* $\mathbf{Pi}^N(P, Q) = 1$;

(b) *If Q is not logically false in L^N and logically implies P in L^N,*

$$\mathbf{Pi}^N(P, Q) = 1;$$

(c) *If Q is not logically false in L^N and logically implies $\sim P$ in L^N,*

$$\mathbf{Pi}^N(P, Q) = 0;$$

(d) *If Q is not logically false in L^N and does not logically imply P or $\sim P$ in L^N,*

$$\mathbf{Pi}^N(P, Q) = 0.$$

Then $\mathbf{Pi}^N$ *meets each one of requirements D18.1(a1)–(a5).*

Proof: PART 1: $\mathbf{Pi}^N$ meets requirements D18.1(a1)–(a4) by virtue of (a)–(d), D4.1, D4.3–5, D2.7, and D2.9.

PART 2: Let $\mathbf{Pi}^N(P, Q \& R)$ and $\mathbf{Pi}^N(Q, R)$ equal 1. If $\mathbf{Pi}^N(P, Q \& R)$ equals 1, it does so by (a), in which case $Q \& R$ is logically false in L^N, or by (b), in which case $Q \& R$ is not logically false in L^N and $Q \& R$ logically implies P in L^N. Hence if $\mathbf{Pi}^N(P, Q \& R)$ equals 1, then by D4.1 and D4.3–4 $Q \& R$ logically implies P in L^N. If $\mathbf{Pi}^N(Q, R)$, on the other hand, equals 1, it does so by (a) or (b) again, and hence by D4.1 and D4.3–4 R logically implies Q in L^N. But if $Q \& R$ logically implies P in L^N and R logically implies Q in L^N, then by D2.7, D4.1, and D4.4 R logically implies $P \& Q$ in L^N. Hence if R is logically false in L^N, $\mathbf{Pi}^N(P \& Q, R)$ equals 1 by (a); and if R is not logically false in L^N, $\mathbf{Pi}^N(P \& Q, R)$ equals 1 by (b).

PART 3: Let $\mathbf{Pi}^N(P \& Q, R)$ equal 1. If $\mathbf{Pi}^N(P \& Q, R)$ equals 1, it does so by (a) or (b); hence by D4.1 and D4.3–4 R logically implies $P \& Q$ in L^N; and hence by D2.7, D4.1, and D4.4 $Q \& R$ logically implies P in L^N and R logically implies Q in L^N. Hence if $Q \& R$ is logically false in L^N, $\mathbf{Pi}^N(P, Q \& R)$ equals 1 by (a); if $Q \& R$ is not logically false in L^N, $\mathbf{Pi}^N(P, Q \& R)$ equals 1 by (b); if R is logically false in L^N, $\mathbf{Pi}^N(Q, R)$ equals 1 by (a); and if R is not logically false in L^N, $\mathbf{Pi}^N(Q, R)$ equals 1 by (b).

PART 4: By virtue of part 2, part 3, and (a)–(d), $\mathbf{Pi}^N(P \& Q, R)$ equals 1 if and only if $\mathbf{Pi}^N(P, Q \& R)$ and $\mathbf{Pi}^N(Q, R)$ both equal 1, $\mathbf{Pi}^N(P \& Q, R)$ equals 0 if and only if $\mathbf{Pi}^N(P, Q \& R)$ or $\mathbf{Pi}^N(Q, R)$ equals 0, and hence $\mathbf{Pi}^N$ meets requirement D18.1(a5).

T21.2. *Let* $\mathbf{Pi}^N$ *be such that:*

(a)–(c) *Like T21.1(a)–(c);*

(d) *If* Q *is not logically false in* L^N *and does not logically imply* P *or* $\sim P$ *in* L^N, $\mathbf{Pi}^N(P, Q) = 1$.

Then $\mathbf{Pi}^N$ *meets each one of requirements D18.1(a1)–(a5).*

Proof: PART 1: $\mathbf{Pi}^N$ meets requirements D18.1(a1)–(a4) by virtue of (a)–(d), D4.1, D4.3–5, D2.7, and D2.9.

PART 2: 1. Let $\mathbf{Pi}^N(P, Q \& R)$ equal 0. If $\mathbf{Pi}^N(P, Q \& R)$ equals 0, then it does so by (c), and hence $Q \& R$ is not logically false in L^N and logically implies $\sim P$ in L^N. But if $Q \& R$ is not logically false in L^N, then by D2.7, D4.1, and D4.3 R is not logically false in L^N; and if $Q \& R$ logically implies $\sim P$ in L^N, then by D2.7, D4.1, and D4.4

R logically implies $\sim(P \,\&\, Q)$ in L^N. Hence $\mathbf{Pi}^N(P \,\&\, Q, R)$ equals 0 by (b).

2. Let $\mathbf{Pi}^N(Q, R)$ equal 0. If $\mathbf{Pi}^N(Q, R)$ equals 0, then it does so by (c) again, and hence R is not logically false in L^N and logically implies $\sim Q$ in L^N. But if R logically implies $\sim Q$ in L^N, then by D2.7, D4.1, and D4.4 R logically implies $\sim(P \,\&\, Q)$ in L^N. Hence $\mathbf{Pi}^N(P \,\&\, Q, R)$ equals 0 by (c).

PART 3: Let $\mathbf{Pi}^N(P \,\&\, Q, R)$ equal 0. If $\mathbf{Pi}^N(P \,\&\, Q, R)$ equals 0, then it does so by (c), and hence R is not logically false in L^N and logically implies $\sim(P \,\&\, Q)$ in L^N. But by D2.7, D4.1, and D4.4 R logically implies $\sim(P \,\&\, Q)$ in L^N if and only if $Q \,\&\, R$ logically implies $\sim P$ in L^N, and by D2.7, D4.1, and D4.3–4 $Q \,\&\, R$ is logically false in L^N if and only if R logically implies $\sim Q$ in L^N. Hence if R is not logically false in L^N and logically implies $\sim(P \,\&\, Q)$ in L^N, then either $Q \,\&\, R$ is not logically false in L^N and logically implies $\sim P$ in L^N or R is not logically false in L^N and logically implies $\sim Q$ in L^N. Hence $\mathbf{Pi}^N(P, Q \,\&\, R)$ or $\mathbf{Pi}^N(Q, R)$ equals 0 by (c).

PART 4: By virtue of part 2, part 3, and (a)–(d), $\mathbf{Pi}^N(P \,\&\, Q, R)$ equals 0 if and only if $\mathbf{Pi}^N(P, Q \,\&\, R)$ or $\mathbf{Pi}^N(Q, R)$ equals 0, $\mathbf{Pi}^N(P \,\&\, Q, R)$ equals 1 if and only if $\mathbf{Pi}^N(P, Q \,\&\, R)$ and $\mathbf{Pi}^N(Q, R)$ both equal 1, and hence $\mathbf{Pi}^N$ meets requirement D18.1(a5).

T21.3. *Let* $\mathbf{Pi}^N$ *be as in T21.1 and let P and Q be G(W) and G(X), respectively, where G is a one-place predicate of* L^N *and W and X are two individual constants of* L^N *distinct from each other. Then* $\mathbf{Pi}^N(P, Q)$ *and* $\mathbf{Pi}^N(\sim P, Q)$ *both equal 0, and, as a result,* $\mathbf{Pi}^N$ *fails to meet requirement D18.1(a6).*

Proof: By D3.2–3, D4.1, and D4.3–4, Q is not logically false in L^N and does not logically imply P or $\sim P$ in L^N. Hence T21.3 by T21.1(d).

T21.4. *Let* $\mathbf{Pi}^N$ *be as in T21.2, and P and Q be as in T21.3. Then* $\mathbf{Pi}^N(P, Q)$ *and* $\mathbf{Pi}^N(\sim P, Q)$ *both equal 1, and, as a result,* $\mathbf{Pi}^N$ *fails to meet requirement D18.1(a6).*

Proof: Similar to the proof of T21.3, but using T21.2(d) instead of T21.1(d).[29]

So much for (1). I now show, in behalf of (2), that when P is an open sentence of L^N and Q a closed sentence of L^N which is not logically false in L^N, $\mathbf{Pi}^N(P, Q)$ is the sum of the various values of which $\mathbf{Ps}^N(P)$ is susceptible, each multiplied by the inductive probability given Q of $\mathbf{Ps}^N(P)$ taking that value (T21.7–8). But the value of a random function, when the function is susceptible of finitely many values, is often estimated in statistics by means of a like sum, the sum of the various values of which the function is susceptible, each multiplied by the statistical probability of the function taking that value. On the strength of T21.7–8 and (1), I therefore conclude that $\mathbf{Pi}^N(P, Q)$ qualifies as an estimate—made in the light of Q—of $\mathbf{Ps}^N(P)$.[30] To simplify matters, I restrict myself to such open sentences P of L^N as contain free occurrences of a single individual variable of L^N. The same result holds, as the reader may verify, for other open sentences P of L^N.

Two theorems about $\mathbf{Ps}^N(P)$ are first in order.

T21.5. *Let P be an open sentence of L^N; let W be the one individual variable of L^N which is free in P; let P_1^*, P_2^*, $\cdots$, and P_m^* be in some arbitrary order the m ($m \geq 0$) instances of P in L^N which are true in L^N; and for each i from 1 to m let W_i be the individual constant of L^N which occurs in P_i^* at all the places where P contains free occurrences of W.*

(a) *If $m = 0$, then $\mathbf{Ps}^N(P) = 0$;*

(b) *If $m > 0$, then $\mathbf{Ps}^N(P) = \sum_{i=1}^{m} \mathbf{w}^N(W_i)$.*

Proof by D13.2(c), D13.2(a), and D2.6(b).

T21.6. *Let P be as in T21.5; let m ($m \geq 0$) be the number of instances of P in L^N which are true in L^N; and for each N from 1 on let $\mathbf{w}^N(W) = \mathbf{w}^N(X)$, where W and X are any two individual constants of L^N. Then $\mathbf{Ps}^N(P) = m/N$.*

Proof by T21.5.

In view of T21.5, $\mathbf{Ps}^N(P)$ is susceptible of (at most) 2^N values, to wit, 0, the weights in L^N of the N individual constants of L^N, the combined weights in L^N of any two of the constants in question, the combined weights in L^N of any three of the constants in question, $\cdots$, and, finally, the combined weights of all the constants in question. In view of T21.6, $\mathbf{Ps}^N(P)$ is susceptible, when the individual constants of L^N $(N = 1, 2, \cdots)$ are all weighted alike in L^N, of (exactly) $N + 1$ values, to wit, $0/N$, $1/N$, $\cdots$, and N/N.

The following illustration should pave the way for T21.7–8. Let the two individual constants of L^2 be allotted $\frac{1}{4}$ and $\frac{3}{4}$ as their respective weights in L^2 and let $\mathbf{Ps}^2(P)$, where P is an open sentence of L^2 of the kind under discussion, be susceptible as a result of four values, to wit, 0, $\frac{1}{4}$, $\frac{3}{4}$, and 1. It can be shown that $\mathbf{Pi}^2(P, Q)$, where Q is a closed sentence of L^N which is not logically false in L^N, is the sum of four products, namely, 0 times the inductive probability given Q of $\mathbf{Ps}^2(P)$ taking 0 as its value, $\frac{1}{4}$ times the inductive probability given Q of $\mathbf{Ps}^2(P)$ taking $\frac{1}{4}$ as its value, $\frac{3}{4}$ times the inductive probability given Q of $\mathbf{Ps}^2(P)$ taking $\frac{3}{4}$ as its value, and 1 times the inductive probability given Q of $\mathbf{Ps}^2(P)$ taking 1 as its value. Proof is as follows.

By D18.1(c) and D2.6(b)

$$\mathbf{Pi}^2(P, Q) = (\tfrac{1}{4}\cdot\mathbf{Pi}^2(P_1{}^*, Q)) + (\tfrac{3}{4}\cdot\mathbf{Pi}^2(P_2{}^*, Q)),$$

where $P_1{}^*$ and $P_2{}^*$ are the two instances of P in L^2. But by D2.7–9, D4.1, and D4.5 $P_1{}^*$ is logically equivalent in L^2 to $(P_1{}^* \;\&\; P_2{}^*) \vee (P_1{}^* \;\&\; \sim P_2{}^*)$ and $P_2{}^*$ to $(P_1{}^* \;\&\; P_2{}^*) \vee (\sim P_1{}^* \;\&\; P_2{}^*)$. Hence by D18.1(a3), D2.6(b), and D2.4

$$\mathbf{Pi}^2(P, Q) = (\tfrac{1}{4}\cdot\mathbf{Pi}^2((P_1{}^* \;\&\; P_2{}^*) \vee (P_1{}^* \;\&\; \sim P_2{}^*), Q)) + (\tfrac{3}{4}\cdot\mathbf{Pi}^2((P_1{}^* \;\&\; P_2{}^*) \vee (\sim P_1{}^* \;\&\; P_2{}^*), Q)).$$

But by D2.7, D4.1, and D4.4 Q logically implies in L^2 both

$$\sim ((P_1{}^* \;\&\; P_2{}^*) \;\&\; (P_1{}^* \;\&\; \sim P_2{}^*))$$

and

$$\sim ((P_1{}^* \;\&\; P_2{}^*) \;\&\; (\sim P_1{}^* \;\&\; P_2{}^*)).$$

Hence by T18.2(e), D2.6(b), and D2.4

$$\mathbf{Pi}^2(P, Q) = (0\cdot\mathbf{Pi}^2(\sim P_1{}^* \;\&\; \sim P_2{}^*, Q)) + (\tfrac{1}{4}\cdot\mathbf{Pi}^2(P_1{}^* \;\&\; \sim P_2{}^*, Q)) + (\tfrac{3}{4}\cdot\mathbf{Pi}^2(\sim P_1{}^* \;\&\; P_2{}^*, Q)) + (1\cdot\mathbf{Pi}^2(P_1{}^* \;\&\; P_2{}^*, Q)).$$

But by D2.7, D3.2, D2.6, and T21.5, $\sim P_1{}^* \;\&\; \sim P_2{}^*$ is true in L^2 if and only if, neither instance of P in L^2 being true in L^2, $\mathbf{Ps}^2(P)$ takes 0 as its value; $P_1{}^* \;\&\; \sim P_2{}^*$ is true in L^2 if and only if, the first instance of P in L^2 and none other being true in L^2, $\mathbf{Ps}^2(P)$ takes $\frac{1}{4}$ as its value; $\sim P_1^* \,\&\, P_2^*$ is true in L^2 if and only if, the second instance of P in L^2 and none other being true in L^2, $\mathbf{Ps}^2(P)$ takes $\frac{3}{4}$ as its value; and $P_1^* \,\&\, P_2^*$ is true in L^2 if and only if, both instances of P in L^2 being true in L^2, $\mathbf{Ps}^2(P)$ takes 1 as its value. $\mathbf{Pi}^2(P, Q)$ thus proves to be the sum of the aforesaid products.

I assumed above that the two individual constants of L^2 were allotted $\frac{1}{4}$ and $\frac{3}{4}$ as their respective weights in L^2. Imagine instead that they are weighted alike in L^2 and that $\mathbf{Ps}^2(P)$ is susceptible as a result of three values, to wit, 0, $\frac{1}{2}$, and 1. $\mathbf{Pi}^2(P, Q)$ can then be shown by the same reasoning as above to equal

$$(0 \cdot \mathbf{Pi}^2(\sim P_1^* \,\&\, \sim P_2^*, Q)) + (\tfrac{1}{2} \cdot (\mathbf{Pi}^2(P_1^* \,\&\, \sim P_2^*, Q) + \mathbf{Pi}^2(\sim P_1^* \,\&\, P_2^*, Q))) + (1 \cdot \mathbf{Pi}^2(P_1^* \,\&\, P_2^*, Q)),$$

and hence by T18.2(e), D2.6(b), and D2.4 to equal

$$(0 \cdot \mathbf{Pi}^2(\sim P_1^* \,\&\, \sim P_2^*, Q)) + (\tfrac{1}{2} \cdot \mathbf{Pi}^2((P_1^* \,\&\, \sim P_2^*) \vee (\sim P_1^* \,\&\, P_2^*), Q)) + (1 \cdot \mathbf{Pi}^2(P_1^* \,\&\, P_2^*, Q)).$$

But by D2.7–8, D3.2, D2.6, and T21.6, $\sim P_1^* \,\&\, \sim P_2^*$ is true in L^2 if and only if, neither instance of P in L^2 being true in L^2, $\mathbf{Ps}^2(P)$ takes 0 as its value; $(P_1^* \,\&\, \sim P_2^*) \vee (\sim P_1^* \,\&\, P_2^*)$ is true in L^2 if and only if, exactly one instance of P in L^2 being true in L^2, $\mathbf{Ps}^2(P)$ takes $\frac{1}{2}$ as its value; and $P_1^* \,\&\, P_2^*$ is true in L^2 if and only if, both instances of P in L^2 being true in L^2, $\mathbf{Ps}^2(P)$ takes 1 as its value. $\mathbf{Pi}^2(P, Q)$ thus proves to be the sum of three products: 0 times the inductive probability given Q of $\mathbf{Ps}^2(P)$ taking 0 as its value, $\frac{1}{2}$ times the inductive probability given Q of $\mathbf{Ps}^2(P)$ taking $\frac{1}{2}$ as its value, and 1 times the inductive probability given Q of $\mathbf{Ps}^2(P)$ taking 1 as its value.

The first result may be generalized as in T21.7, where (1) the four sentences

$$\sim P_1^* \,\&\, \sim P_2^*, \quad P_1^* \,\&\, \sim P_2^*, \quad \sim P_1^* \,\&\, P_2^*, \quad \text{and} \quad P_1^* \,\&\, P_2^*$$

become

$$R_{0_{\binom{N}{0}}}, R_{1_1}, R_{1_2}, \ldots, R_{1_{\binom{N}{1}}}, \ldots, \text{and } R_{N_{\binom{N}{N}}},$$

and (2) the four values 0, $\frac{1}{4}$, $\frac{3}{4}$, and 1 of $\mathbf{Ps}^2(P)$ become

$$v_{0_{\binom{N}{0}}}, v_{1_1}, v_{1_2}, \ldots, v_{1_{\binom{N}{1}}}, \ldots, \text{and } v_{N_{\binom{N}{N}}}.$$

The second may be generalized as in T21.8, where (1) the three sentences

$\sim P_1^* \,\&\, \sim P_2^*$, $(P_1^* \,\&\, \sim P_2^*) \vee (\sim P_1^* \,\&\, P_2^*)$, and $P_1^* \,\&\, P_2^*$

become

$$S_0, S_1, \cdots, \text{and } S_N,$$

and (2) the three values 0, $\frac{1}{2}$, and 1 of $\mathbf{Ps}^2(P)$ become

$$v_0, v_1, \cdots, \text{and } v_N.$$

T21.7. *Let P be an open sentence of L^N in which a single individual variable of L^N is free; let Q be a closed sentence of L^N which is not logically false in L^N; let $P_1^*, P_2^*, \cdots$, and P_N^* be in some arbitrary order the N instances of P in L^N; for each i from 0 to N let $R_{i_1}, R_{i_2}, \cdots$, and $R_{i_{\binom{N}{i}}}$ be in some arbitrary order the $\binom{N}{i}$ results of inserting '$\sim$' before any $N - i$ of the N sentences $P_1^*, P_2^*, \cdots$, and P_N^* in $(\cdots (P_1^* \,\&\, P_2^*) \,\&\, \cdots) \,\&\, P_N^*$; for each i from 0 to N, each j from 1 to $\binom{N}{i}$, and each k from 1 to N, let $v_{i_{j_k}}$ be 0 if P_k^* is preceded in R_{i_j} by '$\sim$', otherwise the weight in L^N of the k-th individual constant of L^N; and, for each i from 0 to N and each j from 1 to $\binom{N}{i}$, let v_{i_j} be $\sum_{k=1}^{N} v_{i_{j_k}}$.*
Then

$$\mathbf{Pi}^N(P, Q) = \sum_{i=0}^{N} \Big(\sum_{j=1}^{\binom{N}{i}} (v_{i_j} \cdot \mathbf{Pi}^N(R_{i_j}, Q))\Big).$$

T21.8. *Let $P, Q, P_1^*, P_2^*, \cdots, P_N^*, R_{0_{\binom{N}{0}}}, R_{1_1}, R_{1_2}, \cdots, R_{1_{\binom{N}{1}}}, \cdots$, and $R_{N_{\binom{N}{N}}}$ be as in T21.7; for each i from 0 to N let S_i be $(\cdots (R_{i_1} \vee R_{i_2}) \vee \cdots) \vee R_{i_{\binom{N}{i}}}$ and v_i be i/N; and for each N from 1 on let $\mathbf{w}^N(W) = \mathbf{w}^N(X)$, where W and X are any two individual constants of L^N. Then*

$$\mathbf{Pi}^N(P, Q) = \sum_{i=0}^{N} (v_i \cdot \mathbf{Pi}^N(S_i, Q)).$$

Proofs for T21.7–8 may be reconstructed from the material of the last two paragraphs or borrowed from Carnap's *Logical Foundations of Probability*.[31]

Since by D2.7–8, D3.2, D2.6, and T21.5–6, (1) R_{i_j} is true in L^N if and only if $\mathbf{Ps}^N(P)$ takes v_{i_j} as its value, and (2) S_i is true in L^N if and only if $\mathbf{Ps}^N(P)$ takes v_i as its value, it follows from T21.7–8 that $\mathbf{Pi}^N(P, Q)$ is the sum of the various values of which $\mathbf{Ps}^N(P)$ is susceptible, each multiplied by the inductive probability given Q of $\mathbf{Ps}^N(P)$ taking that value.

Q has been presumed so far not to be logically false in L^N. Assume, however, that Q is logically false in L^N. By D4.1 and D4.3–4 Q will logically imply in L^N any sentence of L^N you may wish and hence convey any information you may wish. But if Q conveys any information you may wish, then $\mathbf{Pi}^N(P, Q)$ may pass as an estimate—made in the light of Q—of any function of P you may wish, the function $\mathbf{Ps}^N(P)$ itself included. I therefore maintain that $\mathbf{Pi}^N(P, Q)$ qualifies in either case as an estimate—made in the light of Q—of $\mathbf{Ps}^N(P)$.

Now for a caveat: There are, as I pointed out earlier, $2^{\aleph_0}$ functions $\mathbf{w}^N$ which meet requirements D20.5(b1) (or D20.5(b1′)) and D20.5(b2). There are accordingly $2^{\aleph_0}$ functions $\mathbf{Pi}^N$ which meet requirements D18.1(a1)–(a6) (or T19.3(a)–(f)) and whose respective values for a pair of closed sentences P and Q of L^N rate "fair or better" as estimates—made in the light of Q—of the truth-value of P in L^N. I do not, however, claim (let me repeat) that each and every one of those values rates better than "fair" as an estimate—made in the light of Q—of the truth-value of P in L^N. The situation we run into here is like the one which confronted us in Chapter 2, and it is equally critical. How to weight the individual constants of L^N (or, if you prefer, the members of a finite probability set PS) and how to weight the state-descriptions of L^N may indeed be the two most pressing issues in probability theory today.

I shall not attempt in these pages to resolve the second issue.[32] I shall merely list a few avenues along which it can be approached.

One is our inferring habits. We perform inductive inferences of various kinds every day and, though drawing false conclusions many a time, doggedly stick by our ways of doing things. We might therefore codify our inferring habits into so many rules, each one instructing us whether or not to draw a given conclusion from a given premise; compute the values of each function $\mathbf{Pi}^N$ for the ensuing pairs of sentences; and measure through those values the support which $\mathbf{Pi}^N$ lends to the rules in question.[33]

Another avenue is our betting habits. Every day (or nearly every day) we bet on various conjectures and, though losing our stakes many a time, doggedly stick by our ways of doing things. But inductive probabilities may serve—I shall point out in Section 23—as betting quotients. We might therefore codify our betting habits into so many rules, each one instructing us how much to bet on one sentence given another sentence; compute the values of each function $\mathbf{Pi}^N$ for the ensuing pairs of sentences; and measure through those values the support which $\mathbf{Pi}^N$ lends to the rules in question.

A third avenue is current statistical practice. I explained in Section 11 how statisticians, when in the dark as to the probability of a subset A of a probability set PS, draw from PS a sample of the largest size s they can afford, ascertain the number, call it s_1, of members of A in the sample, and estimate the probability of A to be s_1/s. But open sentences play here the same part that sets play with statisticians, and inductive probabilities are meant to serve as estimates of statistical ones. We might therefore compute the values of each function $\mathbf{Pi}^N$ for various pairs of sentences, one an open sentence of which we know the truth-value of s instances, the other a closed sentence to the effect that s_1 of the s instances are true; compare those values with s_1/s; and when the values deviate from s_1/s, list and weigh the factors which deflect them.

Some spadework has already been done towards codifying our inferring habits,[34] some towards codifying our betting ones,[35] and some towards matching suitable values of some of the functions $\mathbf{Pi}^N$ against the above ratio s_1/s.[36] The results, however, are still meager and their significance remains to be assessed with any finality.

22. LOGICAL FALSEHOODS AND LOGICAL TRUTHS QUA EVIDENCE SENTENCES

As the reader may recall, D18.1(a6) read: *If Q is not logically false in L^N, then* $\mathbf{Pi}^N(\sim P, Q) = 1 - \mathbf{Pi}^N(P, Q)$. The restriction on Q was not idle. Imagine indeed that D18.1(a6) were strengthened to read:

D18.1. (a6′) $\mathbf{Pi}^N(\sim P, Q) = 1 - \mathbf{Pi}^N(P, Q)$.

T18.2. (e) could then be strengthened to read:

T18.2. (e′) *If R logically implies $\sim (P \,\&\, Q)$ in L^N, then*

$$\mathbf{Pi}^N(P \vee Q, R) = \mathbf{Pi}^N(P, R) + \mathbf{Pi}^N(Q, R).$$

But by D2.11(a), D4.1, and D4.4 $\mathsf{D}(W, W)$, where W is an individual constant of L^N, logically implies $\sim (P \,\&\, \sim P)$ in L^N. Hence $\mathbf{Pi}^N(P \vee \sim P, \mathsf{D}(W, W))$ would equal $\mathbf{Pi}^N(P, \mathsf{D}(W, W)) + \mathbf{Pi}^N(\sim P, \mathsf{D}(W, W))$. But by D2.11(a), D4.1, and D4.3 $\mathsf{D}(W, W)$ is logically false in L^N. Hence by T18.2(c) $\mathbf{Pi}^N(P, \mathsf{D}(W, W))$ would equal 1, $\mathbf{Pi}^N(\sim P, \mathsf{D}(W, W))$ would equal 1, and $\mathbf{Pi}^N(P \vee \sim P, \mathsf{D}(W, W))$ would equal both 1 and 2, a flat contradiction.

Restrictions of the kind: *If Q is not logically false in L^N*, as in D18.1(a6), or of the kind: *If Q is not logically false in any sublanguage of L^∞ from a given L^N on*, as in T18.5(f), become rather irksome in the long run. Many writers accordingly prefer to bar logical falsehoods from serving as so-called evidence sentences.[37] The matter could be formally handled here as follows. A pair of closed sentences P and Q of a sublanguage L^N of L^∞ could first be said to constitute a probability pair of closed sentences of L^N if Q is not logically false in L^N. The opening lines of D18.1(a) could next be amended to read: *A function* $\mathbf{Pi}^N$ *taking probability pairs of closed sentences of L^N as its arguments, etc.*, and D18.1(a6) amended to read like D18.1(a6′) above. The closing lines of D18.1(c) could finally be amended to read: *Then*

$$\mathbf{Pi}^N(P, Q) = \sum_{i=1}^{N^n} (\mathbf{w}^N(W_{i_1} W_{i_2} \cdots W_{i_n}) \cdot \mathbf{Pi}^N(P_i, Q))$$

if $\mathbf{Pi}^N(P_i, Q)$ *has a value for each i from 1 to N; otherwise* $\mathbf{Pi}^N(P, Q)$ *has no value.* Since a closed sentence Q of L^∞, when logically false in

L^∞, is logically false as well in every sublanguage of L^∞ from a given L^N on, the logical falsehoods of L^∞ as well as of L^1, L^2, L^3, and so on, would be barred by the resulting version of D18.1 from serving as evidence sentences.

As matters stand, **Pi**(P, Q), where Q is logically false in L, can be shown by D4.1, D4.3, T18.2(c), and D18.1(b)–(d) to equal 1 for every closed or open sentence P of L.[38] The result is quite welcome in view of my interpretation of **Pi**(P, Q). As suggested in Section 21, a logical falsehood of L conveys any information one may wish, and hence any information one may need to estimate at 1 the truth-value in L of any sentence of L (so long, of course, as one chooses to estimate that truth-value in the light of a logical falsehood of L).

Oddly enough, **Pi**(P, Q), where Q is logically false in L, would be free to roam at the mercy of P over the entire interval $[0, 1]$ if D18.1(a6) were strengthened to read like D18.1(a6′) above and if D18.1(a2) were weakened to read (for consistency's sake):

D18.1. (a2′) *If P is not logically false in L, then* **Pi**^{N} $(P, P) = 1$.[39]

In view of my interpretation of **Pi**(P, Q), the result would be less welcome than T18.2(c).

Let me pass on, though, to the logical truths of L, which will undoubtedly be frowned upon as evidence sentences. As matters stand, **Pi**(P, Q), where Q is logically true in L, has a value for many a closed and many an open sentence P of L and that value may vary with P, even though Q may be said to convey no information whatever. I, for one, would be willing to estimate the truth-value in L of a sentence P of L in the light of a logical truth of L, say $W = W$, and let my estimate of that truth-value vary with P. To illustrate the second point, I would be willing to estimate in the light of $W = W$ the truth-value in L of a sentence $P_1 \vee P_2$ of L at a normally higher figure than I would the truth-value in L of P_1 alone or of P_2 alone; I would be willing to estimate in that light the truth-value in L of a sentence P_1 & P_2 of L at a normally lower figure than I would the truth-value in L of P_1 alone or of P_2 alone; and so on. Others, however, might balk at making such estimates. The logical truths of L also prove to be troublesome as evidence sentences when

Pi(P, Q) is interpreted—as is often done—as the degree to which P is confirmed in L by Q or given Q.[40] I therefore submit two alternatives to the course that I officially favor here.

(1) The reader might retain my interpretation of **Pi**(P, Q) but bar the logical truths of L from serving as evidence sentences. The matter could be formally handled as follows. A pair of closed sentences P and Q of a given sublanguage L^N of L^∞ could first be said to constitute a probability pair of closed sentences of L^N if Q is not logically true in L^N. The opening lines of D18.1(a) and the closing ones of D18.1(c) could then be amended to read as in the second paragraph of this section.

(2) Or the reader might, if he so preferred, retain the logical truths of L as evidence sentences but modify slightly my interpretation of **Pi**(P, Q). As matters stand, the truth-value in L of a sentence P of L is to be estimated in the light of the information conveyed by a closed sentence Q of L and in the light of that information only. Note, however, that in everyday life the truth-value of a sentence is often estimated in the light of two things: first, some information which bears directly on the sentence and, for that reason, may be appended to the estimate; second, some more general information which may bear on other sentences as well and is tacitly taken for granted. Note also that D18.1(a) yields the following consequences, where any closed sentence of L^N, and hence any closed sentence of L^N which is not logically true in L^N, can serve as R in (a), as Q in (b), as S in (c)–(e), and (under stated circumstances) as R in (f):

T22.1. (a) $0 \leq \mathbf{Pi}^N(P, Q \,\&\, R)$;

(b) $\mathbf{Pi}^N(P, P \,\&\, Q) = 1$;

(c) *If P and Q are logically equivalent in L^N, then*

$$\mathbf{Pi}^N(P, R \,\&\, S) = \mathbf{Pi}^N(Q, R \,\&\, S);$$

(d) *If Q and R are logically equivalent in L^N, then*

$$\mathbf{Pi}^N(P, Q \,\&\, S) = \mathbf{Pi}^N(P, R \,\&\, S);$$

(e) $\mathbf{Pi}^N(P \,\&\, Q, R \,\&\, S) = \mathbf{Pi}^N(P, (Q \,\&\, R) \,\&\, S) \cdot \mathbf{Pi}^N(Q, R \,\&\, S)$;

(f) *If $Q \,\&\, R$ is not logically false in L^N, then*

$$\mathbf{Pi}^N(\sim P, Q \,\&\, R) = 1 - \mathbf{Pi}^N(P, Q \,\&\, R).$$ [41]

Some closed sentence Q' of L^∞ such that Q' is not logically true in L^∞ nor in any sublanguage of L^∞ from a given L^M on, could accordingly be presumed to be on hand and, if so wished, to convey the general information in the light of which the truth-value in L of any sentence P of L^∞ or of any sublanguage of L^∞ from L^M on is to be partly or wholly estimated. A pair of closed sentences P and Q of a given sublanguage $L^N (N \geq M)$ of L^∞ could next be said to constitute a probability pair of closed sentences of L^N if $Q \,\&\, Q'$ is not logically false in L^N. The opening lines of D18.1(a) could next be amended to read: *A function* $\mathbf{Pi}^N (N \geq M)$ *taking probability pairs of closed sentences of* L^N *as its arguments, etc.*, and D18.1(a6) amended to read like D18.1(a6′) on page 122.[42] The closing lines of D18.1(c) could finally be amended to read as in the second paragraph of this section. This done, the reader might reinterpret $\mathbf{Pi}(P, Q)$ as an estimate—made in the light of the information (if any) conveyed by Q and that conveyed by Q'—of the truth-value of P in L and, hence, when Q is logically true in L, as an estimate—made in the light of the information conveyed by Q' alone—of the truth-value of P in L.

Under alternative (1) absolute inductive probabilities could still be allotted to the closed sentences of L^N; they would be mere expedients, though, for reckoning conditional inductive probabilities. With $\mathbf{Pi}_0{}^N$ meeting all five of requirements D19.1(a)–(e), the latter probabilities could be reckoned as follows:

D19.2. (a′) *Let P and Q be two closed sentences of L^N.*
(a1′) *If Q is logically true in L^N, then $\mathbf{Pi}^N(P, Q)$ has no value;*
(a2′) *If Q is neither logically true nor logically false in L^N, then*

$$\mathbf{Pi}^N(P, Q) = \mathbf{Pi}_0{}^N(P \,\&\, Q)/\mathbf{Pi}_0{}^N(Q);$$

(a3′) *If Q is logically false in L^N, then*

$$\mathbf{Pi}^N(P, Q) = 1.$$

(b)–(d) *Like D18.1(b)–(d).*

The resulting function $\mathbf{Pi}^N$ would meet all seven of requirements D18.1(a1)–(a7). With $\mathbf{Pi}_0{}^N$ meeting only requirements D19.1(a)–(d), on the other hand, conditional inductive probabilities could be reckoned as follows:

D19.2. (a″) *Let P and Q be two closed sentences of* L^N.

(a1″) *If Q is logically true in* L^N, *then* $\mathbf{Pi}^N(P, Q)$ *has no value;*

(a2″) *If Q is not logically true in* L^N *and* $\mathbf{Pi}_0{}^N(Q) \neq 0$, *then*

$$\mathbf{Pi}^N(P, Q) = \mathbf{Pi}_0{}^N(P \,\&\, Q)/\mathbf{Pi}_0{}^N(Q);$$

(a3″) *If Q is not logically true in* L^N *and* $\mathbf{Pi}_0{}^N(Q) = 0$, *then*

$$\mathbf{Pi}^N(P, Q) = 1.$$

(b)–(d) *Like D18.1(b)–(d).*

The resulting function would meet only requirements T19.3(a)–(f).

Under alternative (2) absolute inductive probabilities could likewise be allotted to the closed sentences of $L^N (N \geq M)$; $\mathbf{Pi}_0{}^N(P)$ would now prove to be an estimate—made in the light of the information conveyed by Q' alone—of the truth-value of P in L. $\mathbf{Pi}_0{}^N$, however, could no longer be expected to meet requirement D19.1(e) nor $\mathbf{Pi}^N$ to meet requirement D18.1(a7). Suppose indeed that $\mathbf{Tv}^N(P)$ were estimated in the light of Q' at 1. Reasonable though it might be under the circumstances to insist on Q' logically implying P in L^N, one could hardly insist on P being logically true in L^N. Or suppose that $\mathbf{Tv}^N(P)$ were estimated in the light of $Q \,\&\, Q'$ at 1. Reasonable though it might be to insist on $Q \,\&\, Q'$ logically implying P in L^N, one could hardly insist on Q alone doing so. Conditional inductive probabilities, as a result, would have to be reckoned as follows:

D19.2. (a‴) *Let P and Q be two closed sentences of* L^N $(N \geq M)$.

(a1‴) *If* $Q \,\&\, Q'$ *is logically false in* L^N, *then* $\mathbf{Pi}^N(P, Q)$ *has no value;*

(a2‴) *If* $Q \,\&\, Q'$ *is not logically false in* L^N *and* $\mathbf{Pi}_0{}^N(Q) \neq 0$, *then*

$$\mathbf{Pi}^N(P, Q) = \mathbf{Pi}_0{}^N(P \,\&\, Q)/\mathbf{Pi}_0{}^N(Q);$$

(a3‴) *If* $Q \,\&\, Q'$ *is not logically false in* L^N *and* $\mathbf{Pi}_0{}^N(Q) = 0$, *then*

$$\mathbf{Pi}^N(P, Q) = 1.$$

(b)–(d) *Like D18.1(b)–(d).*

The resulting function $\mathbf{Pi}^N$ would meet only requirements T19.3(a)–(f).

23. PERSONAL VERSUS INDUCTIVE PROBABILITIES

A bet is a contract under which one person agrees to pay another a given sum, call it u_1, if a given closed sentence P proves to be false, and to collect from the other person another sum, call it u_2, if P proves to be true. The first person is said, under such circumstances, to be betting on P with odds of u_1 to u_2 or, as the matter is sometimes put, with a betting quotient equal to $u_1/(u_1 + u_2)$, the second to be betting on $\sim P$ with odds of u_2 to u_1 or with a betting quotient equal to $u_2/(u_1 + u_2)$. It has been proposed by Ramsey, de Finetti, and others, that the highest betting quotient with which a person would bet on a closed sentence P be termed the probability allotted by that person to P and, by extension, that the highest betting quotient with which a person would bet on a closed sentence P in the light of the information conveyed by another closed sentence Q be termed the probability allotted by that person to the pair of sentences P and Q.[43]

Personal probabilities, as probabilities of the above sort have come to be called, may differ upon occasion from inductive probabilities. Followers of Keynes hold, to be sure, that inductive probabilities qualify as betting quotients, and no one, so far as I know, has seriously disputed their claim. There is little doubt, however, that, suitable though inductive probabilities may be as betting quotients, the average gambler all too frequently uses figures at variance with them when offering or accepting odds in everyday life. Inductive probabilities meet by definition requirements D18.1(a)–(b) or analogues thereof; personal probabilities, on the other hand, frequently violate those requirements. The average gambler might fail, for example, to realize that two closed sentences P and P' are logically equivalent, and offer in the light of some closed sentence Q higher odds on P than he would on P', thus violating requirement D18.1(a3); he might also fail to realize that two closed sentences Q and Q' are logically equivalent, and offer in the light of Q higher odds on some closed sentence P than he would in the light of Q', thus violating requirement D18.1(a4); and so on.

Personal probability theorists are not blind to such discrepancies between the figures we should use when offering or accepting odds and the figures we happen to use. They soon draw up as a result a list of requirements which, as I understand it, personal probabilities should meet, dismiss as unfair or incoherent those which do not meet them, and address themselves exclusively to those that do.[44] The list is not the same in all writers. Patrick Suppes, for example, places upon personal probabilities requirements which are somewhat weaker than D18.1(a)–(b).[45] Most writers, however, call upon D18.1(a)–(b) or analogues thereof as their favored requirements. Though personal probabilities may differ upon occasion from inductive probabilities, those among them that commonly pass muster as fair or coherent are thus nothing but inductive probabilities of a sort.

Savage has recently offered to another though related account of personal probabilities.[46] He presumes given a set $\vee_P$ of logical possibilities, subsets of which he refers to as events. He also presumes that, for each pair of subsets A and B of $\vee_P$, a given subject is offered a choice between: (1) winning a given prize p_A or winning a given prize p_B according as A takes place or not, and (2) winning prize p_A or winning prize p_B according as B takes place or not. He then proposes that A be said to be not more probable to his subject than B if the subject prefers prize p_A to prize p_B and yet does not choose (1) over (2). Suppose, to use Savage's own illustration, that two eggs, a brown one and a white one, are about to be broken; suppose also that a given subject is offered a choice between (1′) winning one dollar or winning nothing according as the brown egg turns out to be fresh or not, and (2′) winning the dollar or winning nothing according as the white egg turns out to be fresh or not; suppose finally that the subject though preferring a dollar to nothing, does not choose (1′) over (2′). Savage would then take it that to the subject in question the chances of the brown egg being fresh do not exceed the chances of the white one being fresh.

Savage's proposal is easily amended to suit the present context. The state-descriptions of a given sublanguage L^N of L^∞, for example, can do duty for Savage's logical possibilities, and the closed sentences of L^N do duty for Savage's events. For each pair of closed sentences P and Q of L^N Savage's subject can then be offered the choice between (1″) winning a given prize p_P or winning a given prize p_Q according as P proves to be true in L_N or not, and (2″) winning prize p_P or winning prize p_Q according as Q proves to be true in L^N or not. This done, Savage's subject can finally be said to rate P not more probable than Q in L^N, if he prefers prize p_P to prize p_Q and yet does not choose (1″) over (2″).

Savage goes on to show that under certain circumstances a function taking subsets of $\vee_P$ as its arguments, taking real numbers as its values, and meeting requirements analogous to D19.1(a)–(d), can be constructed on the basis of his comparative probability function. A twin function, taking closed sentences of L^N as its arguments, taking real numbers as its values, and meeting requirements D19.1(a)–(d), can likewise be constructed on the basis of the comparative probability relation I have just defined.

The comparative probability assessments, P is not more probable than Q in L^N, P' is not more probable than Q' in L^N, and so on, arrived at in the course of such a lottery as the above, might well differ from the corresponding inequalities $\mathbf{Pi}_O{}^N(P) \leq \mathbf{Pi}_O{}^N(Q)$, $\mathbf{Pi}_O{}^N(P') \leq \mathbf{Pi}_O{}^N(Q')$, and so on. Writers of Savage's persuasion soon restrict themselves, however, to such assessments as match, mutatis mutandis, the inequalities $\mathbf{Pi}_O{}^N(P) \leq \mathbf{Pi}_O{}^N(Q)$, $\mathbf{Pi}_O{}^N(P') \leq \mathbf{Pi}_O{}^N(Q')$, and so on.[47] To all intents and purposes, personal qualitative probabilities, as Savage's assessments have come to be called, are thus nothing but inductive inequalities of a sort.

I would accordingly conclude, in the light of the above analysis, that personal probabilities, be they of the Ramsey-de Finetti sort or of the Savage sort, soon turn in their sponsors' very hands into inductive probabilities or inductive inequalities of a sort. The study of our betting habits which Ramsey, de Finetti, and Savage have undertaken may, however, suggest further requirements to impose upon inductive probability functions and hence help, as I noted before, to weed out some of the all too numerous functions which qualify at the moment as inductive probability functions.

24. A SECOND LOOK AT INDUCTIVE INFERENCES

I close this book with a few additional remarks on inductive inferences.

Statisticians, as the reader may recall from Section 17, would normally trust a member of a probability set PS to belong to a subset A of PS if the member of PS in question is known to belong to another subset B of PS and the probability of A given B is very high or, to be more realistic about it, is estimated to be very high. Given two closed sentences P and Q of L, one might likewise allow Q to be inductively inferred as a conclusion from P as a premise if P is known to be true in L and if $\mathbf{CRs}(Q, P)$, the coefficient of statistical reliability of the inference in question, is estimated to be very high. But by virtue of D13.3(a) and T21.7–8, (1) $\mathbf{CRs}(Q, P)$ is equal to $\mathbf{Ps}(Q' \,\&\, P')/\mathbf{Ps}(P')$, where Q' and P' are the individual-constant-free mates in L of Q and P, respectively, and (2) $\mathbf{Pi}(Q' \,\&\, P', W = W)$ and $\mathbf{Pi}(P', W = W)$, where W is an individual constant of L, constitute estimates (made in the light of $W = W$) of $\mathbf{Ps}(Q' \,\&\, P')$ and $\mathbf{Ps}(P')$, respectively. If willing in view of (1) and (2) to treat $\mathbf{Pi}(Q' \,\&\, P', W = W)/\mathbf{Pi}(P', W = W)$ as an estimate of $\mathbf{CRs}(Q, P)$, one might therefore allow Q to be inductively inferred from P if P is known to be true in L and if $\mathbf{Pi}(Q' \,\&\, P', W = W)/\mathbf{Pi}(P', W = W)$ is very high (Rule R1).

Other rules of inference come to mind, which, though not dictated like the above one by current statistical practice, may nevertheless deserve consideration. Given again two closed sentences P and Q of L, one might, for example, allow Q to be inductively inferred from P over $\sim Q$ if P is known to be true in L, if $\mathbf{Pi}(Q, P)$ is larger than $\mathbf{Pi}(\sim Q, P)$, and if no closed sentence R of L such that $\mathbf{Pi}(Q, P \,\&\, R) \neq \mathbf{Pi}(Q, P)$ is known to be true in L.[48] Or, given a closed sentence P of L and n ($n \geq 2$) closed sentences $Q_1, Q_2, \cdots$, and Q_n of L such that at least one of $Q_1, Q_2, \cdots$, and Q_n must be true in L and at most one of $Q_1, Q_2, \cdots$, and Q_n can be true in L,[49] one might allow Q_i ($i \leq n$) to be inductively inferred from P over $Q_1, Q_2, \cdots, Q_{i-1}, Q_{i+1}, \cdots$, and Q_n if P is known to be true in L, if $\mathbf{Pi}(Q_i, P)$ is the largest one of $\mathbf{Pi}(Q_1, P)$, $\mathbf{Pi}(Q_2, P)$, $\cdots$, and $\mathbf{Pi}(Q_n, P)$, and if no closed sentence R of L such that $\mathbf{Pi}(Q_j, P \,\&\, R) \neq \mathbf{Pi}(Q_j, P)$ is known, for any j from 1 to n, to be true in L (Rule R2).

Inductive inferences cannot, in general, be adjudged permissible or not by means of R1 or R2 unless a specific function $\mathbf{Pi}^N$ has been settled on and, in the case of R1, specific weights have been allotted to the individual constants of L^N for each N from 1 on. Many a reader will doubtless balk at paying such a fee for availing himself of either rule. There are fortunately two major kinds of inductive inferences whose merits can be assessed by means of R2 at nearly bargain rates.

Two theorems are first in order.

In *Logical Foundations of Probability* Carnap showed that a version of the so-called hypergeometric theorem holds for any function $\mathbf{Pi}^N$ which is both strongly regular and symmetrical. His version of the theorem may be made to read:

T24.1. (1) *Let* $W_1, W_2, \cdots,$ *and* W_p *be* p $(p \geq 1)$ *individual constants of* L^N *distinct from one another;*

(2) *let* $W_{i_1}, W_{i_2}, \cdots,$ *and* W_{i_s} *be any distinct* s $(s \geq 1)$ *of the* p *individual constants* $W_1, W_2, \cdots,$ *and* W_p*;*

(3) *let* $P_1, P_2, \cdots,$ *and* $P_{\binom{p}{p_1}}$ *be in some arbitrary order the* $\binom{p}{p_1}$ *results of inserting '~' before any* $p - p_1$ $(0 \leq p_1 \leq p)$ *of the* p *sentences* $G(W_1), G(W_2), \cdots,$ *and* $G(W_p)$ *in* $(\cdots (G(W_1) \mathbin{\&} G(W_2)) \mathbin{\&} \cdots) \mathbin{\&} G(W_p)$*;*

(4) *let* P *be* $(\cdots (P_1 \vee P_2) \vee \cdots) \vee P_{\binom{p}{p_1}}$*;*

(5) *let* $Q_1, Q_2, \cdots,$ *and* $Q_{\binom{s}{s_1}}$ *be in some arbitrary order the* $\binom{s}{s_1}$ *results of inserting '~' before any* $s - s_1$ $(max(0, s - p + p_1) \leq s_1 \leq min(s, p_1))$ *of the* s *sentences* $G(W_{i_1}), G(W_{i_2}), \cdots,$ *and* $G(W_{i_s})$ *in* $(\cdots (G(W_{i_1}) \mathbin{\&} G(W_{i_2})) \mathbin{\&} \cdots) \mathbin{\&} G(W_{i_s})$*;*
and

(6) *let* Q *be* $(\cdots (Q_1 \vee Q_2) \vee \cdots) \vee Q_{\binom{s}{s_1}}$.

Then

$$\mathbf{Pi}^N(Q, P) = \frac{\binom{p_1}{s_1}\binom{p - p_1}{s - s_1}}{\binom{p}{s}}.$$ [50]

The sentence referred to as P in T24.1 says—roughly—that p_1 of the p individuals designated in L^N by $W_1, W_2, \cdots,$ and W_p belong to the extension in L^N of the one-place predicate G of L^N; the one referred to as Q says—roughly again—that s_1 of the s individuals designated in L^N by some s or other of the self-same constants $W_1, W_2, \cdots$, and W_p belong to the extension of G in L^N; and T24.1 sets the inductive probability in L^N of the second sentence given the first at the familiar figure

$$\frac{\binom{p_1}{s_1}\binom{p-p_1}{s-s_1}}{\binom{p}{s}}.$$

Improving upon Carnap's result, I have recently obtained a second version of the hypergeometric theorem holding, more generally, for any function $\mathbf{Pi}^N$ which is regular. My version of the theorem is closer word-for-word to Theorem I′ on page 56. It presupposes, though, like its counterpart on page 56, that the sample whose composition it deals with is drawn at random from its parent population, a notion formally defined in D24.2.[51]

D24.2. *Let A, B, B_1, and B_2 contain no individual constant of L.*[52]

(a) *The sample designated in a given sublanguage L^N of L^∞ by A is said to be drawn at random from the population designated in L^N by B with respect to a given partition of the population in question into two cells respectively designated in L^N by B_1 and B_2 if*

$$\mathbf{Pi}^N(X_1, X_2, \cdots, X_s \in A, (B = B_1 \cup B_2 \,\&\, (B_1 = \{W_1, W_2, \cdots, W_{p_1}\} \,\&\, B_2 = \{W_{p_1+1}, W_{p_1+2}, \cdots, W_p\})) \,\&\, (A \subset B \,\&\, S(A) = s)) = 1/\binom{p}{s},$$

where p is any integer from 1 to N, p_1 is any integer from 0 to p, s is any integer from 1 to p, $W_1, W_2, \cdots$, and W_p are any p individual constants of L^N distinct from one another, and $X_1, X_2, \cdots$, and X_s are any distinct s of the p individual constants $W_1, W_2, \cdots$, and W_p.

(b) *The sample designated in L^∞ by A is said to be drawn at random from the population designated in L^∞ by B with respect to a given partition of the population in question into two cells respectively designated in L^∞ by B_1 and B_2 if for every N from 1 on the sample designated in the sublanguage L^N of L^∞ by A is drawn at random from the population designated in L^N by B with respect to the partition of the population in question into the two cells respectively designated in L^N by B_1 and B_2.*

T24.3. (1) *Let A, B, B_1, and B_2 contain no individual constant of L;*

(2) *let the sample designated in L by A be drawn at random from the population designated in L by B with respect to a given partition of the population in question into the two cells respectively designated in L by B_1 and B_2;*
and

(3) *let P be*

$$((B = B_1 \cup B_2 \,\&\, B_1 \not{o}\, B_2) \,\&\, (S(B) = p \,\&\, S(B_1) = p_1)) \,\&\, (A \subset B \,\&\, S(A) = s)$$

and Q be

$$S(A \cap B_1) = s_1,$$

where $p \geq 1$, $0 \leq p_1 \leq p$, $1 \leq s \leq p$, *and* $max(0, s - p + p_1) \leq s_1 \leq min(s, p_1)$.
Then

$$\mathbf{Pi}(Q, P) = \frac{\binom{p_1}{s_1}\binom{p - p_1}{s - s_1}}{\binom{p}{s}}.$$ [53]

The statistical counterpart of T24.3 had to do, the reader may recall, with an unspecified one of the possible outcomes of drawing an (*a*)-sample or a (*b*1)-sample from a finite population; T24.3, on the other hand, has to do with the actual outcome of drawing an (*a*)-sample from such a population. Otherwise the two theorems match exactly.

Now for some of the uses to which the two results may be put.

Inductive inferences whose premise and conclusion respectively detail (like the above *P* and *Q*) the composition of a population and that of a sample drawn from the population, have come to be called direct or population-to-sample inferences. In view of the above results, direct inferences can be appraised by means of R2 for a very small fee. So long in fact as we will opt for functions $\mathbf{Pi}^N$ that are both strongly regular and symmetrical or, if so preferred, merely opt for functions $\mathbf{Pi}^N$ that are regular and presume (as statisticians so frequently end up doing) that the sample on hand has been drawn at random from its parent population, the reader can read from a table or chart for the hypergeometric distribution many—and, upon occasion, all—of the figures he may need to appraise a direct inference by means of R2.

Inductive inferences whose premise and conclusion respectively detail the composition of a sample and that of the population from which the sample has been drawn, have come to be called, on the other hand, inverse or sample-to-population inferences. It was long believed, and is still believed by many, that inverse inferences can be appraised only after specific weights have been allotted to the state-descriptions of L^N. In the nineteen-twenties, however, Jerzy Neyman and Egon S. Pearson showed that the corresponding inferences with sets or, equivalently, with open sentences in place of closed ones can be appraised by means of suitable direct inferences.[54] The Neyman-Pearson method for appraising inverse inferences being readily adapted to the present context,[55] I beg leave to conclude that by means of R2 inverse as well as direct inferences can be appraised for a very small fee.

NOTES

[1] Or equivalents of those requirements. The 'i' in '$\mathbf{Pi}^N$' is short of course for 'inductive.' '$\mathbf{Pi}^N(P, Q)$' may thus be read 'The inductive probability in L^N of P given Q,' and '$\mathbf{Pi}^\infty (P, Q)$' one line below in the text be read 'The inductive probability in L^∞ of P given Q.'

[2] When neither one of P and Q contains individual constants, L^N is L^1; when either one does, L^N is the first sublanguage of L^∞ among whose primitive signs the constants in question figure. The procedure described in the text is used by Carnap in *Logical Foundations of Probability*, §§ 53–54, *The Continuum of Inductive Methods*, pp. 9 and 12, and so on. Writers like I. J. Good, J. Hosiasson, H. Jeffreys, J. M. Keynes, and G. H. von Wright, who operate with a single language (instead of the present family of languages L), also operate with a single inductive probability function (instead of the present family of functions $\mathbf{Pi}^1$, $\mathbf{Pi}^2$, $\mathbf{Pi}^3$, and so on ad infinitum, and $\mathbf{Pi}^\infty$); they place, however, upon that function the same requirements or nearly the same requirements as are placed here upon the functions $\mathbf{Pi}^1$, $\mathbf{Pi}^2$, $\mathbf{Pi}^3$, and so on. Carnap eventually adds to the present list of requirements in *Logical Foundations of Probability*, *The Continuum of Inductive Methods*, and in other writings of his which I list in the Bibliography; for brief comments on the matter, see Sections 20–21.

[3] In "On chances and estimated chances of being true," the author proposed that **Pi**(P, Q), where P is an open sentence and Q a closed sentence of L, be taken to be

$$\sum_{i=1}^{N} \left(\frac{1}{N} \cdot \mathbf{Pi}^N(P_i, Q)\right)$$

or

$$\operatorname*{Limit}_{N\to\infty} \sum_{i=1}^{N} \left(\frac{1}{N} \cdot \mathbf{Pi}^\infty(P_i, Q)\right),$$

where, for each N from 1 on and each i from 1 to N, P_i is like P except for containing occurrences of the i-th individual constant of L^N at all the places where P contains free occurrences of some individual variable of L. The allotments described in the text were proposed by him in his paper "Statistical and Inductive Probabilities."

[4] Or, by virtue of D4.5 and D4.2, *if* $P \equiv Q$ *is logically true in* L^N.

[5] Or, by virtue of D4.5 and D4.2, *if* $Q \equiv R$ *is logically true in* L^N.

[6] Or, by virtue of D4.3 and D4.2, *if* $\sim Q$ *is not logically true in* L^N. The restriction may of course be omitted when Q in $\mathbf{Pi}^N(P, Q)$ is presumed all along not to be logically false in L^N, a procedure adopted by Carnap and others.

[7] D18.1(a1)–(a6) are essentially due to G. H. von Wright, *The Logical Problem of Induction*, pp. 92–93. Various sets of requirements equivalent to or about equivalent to D18.1(a1)–(a6) will be found in the author's "On logically false evidence statements."

[8] When, for each N from 1 on, any two sequences of n ($n \geq 1$) individual constants of L^N are allotted the same weight in L^N, the sum and the limit in question finally reduce to those shown in Note 3 of this chapter.

[9] The proof of (d) is borrowed from K. R. Popper, *The Logic of Scientific Discovery*, pp. 352–353.

[10] The proof of (e) is borrowed from von Wright, *loc. cit.*, p. 94.

[11] That $W = W$ is not logically false in L^N is a corollary by D3.2–3 of the following theorem: *If* P *is logically false in* L^N, *then* P *is false in* L^N.

[12] The proof of (b) was suggested to me by a referee for *The Journal of Symbolic Logic*.

[13] It does not follow, however, that if a closed sentence P of L^∞ is not logically false in L^∞, then P is not logically false either in every sublanguage of L^∞ from a given L^N on. T18.5(f) is therefore not provable here under the form: *If* Q *is not logically false in* L^∞, *then* $\mathbf{Pi}^\infty(\sim P, Q) = 1 - \mathbf{Pi}^\infty(P, Q)$, which makes for a slight lack of symmetry between inductive probabilities in L^N and inductive probabilities in L^∞. For further remarks on inductive probabilities in L^∞, see the author's "The Problem of the Confirmation of Laws."

[14] Like-minded requirements are to be found in F. Waismann, "Logische Analyse des Wahrscheinlichkeitsbegriffs," pp. 236–237. Waismann, however,

operates with a single language rather than the present family of languages L. The '0' in '$\mathbf{Pi}_0{}^N(P)$' is short for 'null,' an epithet often used in lieu of 'absolute.'

[15] Once D19.1(e) is added to D19.1, $\mathbf{Pi}_0{}^N(Q) \neq 0$ in D19.2(a1) may be made to read: *Q is not logically false in L^N*, and $\mathbf{Pi}_0{}^N(Q) = 0$ in D19.2(a2) made to read: *Q is logically false in L^N*, a practice often adopted in the literature.

[16] Carnap has 'regular' where I have 'strongly regular' and has 'quasi-regular' (or, upon occasion, 'non-regular') where I have 'regular'; see *Notes on Probability and Induction*, pp. (3)–(4), and *An Axiom System of Inductive Logic*, pp. 60–71. For further details on D19.1(e), D18.1(a8), and so on, see the author's "On So-called Degrees of Confirmation," pp. 314–315.

[17] Note for proof that: (1) $(\forall W)P$, where W is not free in P, is logically equivalent to P in L^N and hence may be replaced by P wherever it occurs, (2) $(\forall W)P$, where W is free in P, is logically equivalent in L^N to $(\cdots(P_1 \,\&\, P_2) \,\&\, \cdots) \,\&\, P_N$, where, for each i from 1 to N, P_i is like P except for containing occurrences of the i-th individual constant of L^N at all the places where P contains free occurrences of W, and hence may be replaced by $(\cdots(P_1 \,\&\, P_2) \,\&\, \cdots) \,\&\, P_N$ wherever it occurs, (3) $W = W$, where W is an individual constant of L^N, is logically equivalent in L^N to $G(W) \supset G(W)$, where G is any one-place predicate of L^N, and hence may be replaced by $G(W) \supset G(W)$ wherever it occurs, and (4) $W = X$, where W and X are two individual constants of L^N distinct from each other, is logically equivalent in L^N to $\sim (G(W) \supset G(W))$, where G is as in (3), and hence may be replaced by $\sim (G(W) \supset G(W))$ wherever it occurs. Repeated applications of (1)–(4) will turn any closed sentence of L^N into a closed quantifier-and-identity-free sentence of L^N which is logically equivalent to it in L^N.

[18] See the author's "On requirements for conditional probability functions." (1) and (2) are extensions by the author of a result of Popper's, *loc. cit.*, *Appendix *v*. D18.1(a1), D18.1(a2), T18.4(a), T18.4(b), D18.1(a5), and T18.4(c) do not presuppose, the way D18.1(a1)–(a6) do, definitions D4.1–2, and for that reason are called by Popper autonomous requirements.

[19] See Carnap, *Logical Foundations of Probability*, chapter V. For a closely related method of allotting inductive probabilities to closed sentences, see O. Helmer and P. Oppenheim, "A syntactical definition of probability and of degree of confirmation," and C. G. Hempel and P. Oppenheim, "A Definition of "Degree of Confirmation"."

[20] The order in which $Q_1, Q_2, \cdots$, and Q_{α_N} are arranged here does not matter since conjunction, the operation rendered by '&,' is commutative.

[21] See p. 87. In "A logical measure function," where he explicitly deals with inductive probabilities, Kemeny offers a number of amendments to D20.5–7, which for lack of space I cannot recount here.

[22] Carnap favors D20.5(b1) over D20.5(b1′); the latter appears in von Wright, "Carnap's Theory of Probability," p. 368n.

[23] *Loc. cit.*, chapter VIII. In Carnap $\mathbf{Pi}^N(P, Q)$ comes to equal $\mathbf{Pi}^N(P', Q')$, where P and Q are two closed sentences of L^N and P' and Q' are, respectively, like P and Q except for containing occurrences of an individual constant W' of L^N at those and only those places where P and Q contain occurrences of an individual constant W of L^N. Functions $\mathbf{Pi}^N$ of which this holds true are by extension called symmetrical.

[24] The suggestion was made in the author's paper "Statistical and Inductive Probabilities." In an early draft of that paper I placed no restriction on P; this, however, led to contradiction, as Professor Israël Sheffer pointed out to me.

[25] In *The Continuum of Inductive Methods*, pp. 40–44, Carnap studies another allotment which also has some claim on the label 'relative frequency allotment.'

[26] For pertinent references, see Carnap, *Logical Foundations of Probability*, p. 565, and von Wright, *The Logical Problem of Induction*, pp. 210–211.

[27] Truth-values and statistical probabilities are normally estimated in the light of sentences which are either known or presumed to be true. But only closed (as opposed to open) sentences of L can be true in L. Hence the restriction imposed on Q in $\mathbf{Pi}(P, Q)$ throughout this chapter.

[28] The interpretation stems in part from Carnap's treatment of inductive probabilities as estimates of relative frequencies in *Logical Foundations of Probability*, pp. 168–175 and 540–549, and bears some likeness, as Professor Carnap reminded me, to C. I. Lewis' views on probability in *An Analysis of Knowledge and Valuation*, pp. 290–292. Phrased slightly differently in "On chances and estimated chances of being true" and "A New Interpretation of $\mathbf{c}(h, e)$," where I first proposed it, it reached its present form in "Statistical and Inductive Probabilities" and "Probabilities as Truth-Value Estimates." Much of this section is borrowed, by the way, from the latter two papers. Inductive probabilities have also been interpreted as betting quotients (a point I take up in Section 23) and as degrees of confirmation; see Carnap, *loc. cit.*, pp. 162–182, on this whole matter.

[29] The reader may verify on his own that the functions $\mathbf{Pi}^N$ defined in T21.1 and T21.2 also fail to meet requirement T19.3(f). For an inductive probability function which fails to satisfy D18.1(a5), see the Helmer, Hempel, and Oppenheim papers mentioned in Note 19 of this chapter.

[30] The idea of estimating—in the light of a sentence Q—the value of a function f as the sum of the various values of which f is susceptible, each multiplied by the inductive probability given Q of f taking that value, goes back, it would seem, to Carnap, *loc. cit.*, §§ 99–100.

[31] See *loc. cit.*, pp. 540–545. Note that on page 542, line 11, the first occurrence of 'm' should be changed to '$s - m$.'

[32] In compensation I offer in Section 24 some interim ways of appraising various inductive inferences.

[33] One of those rules would undoubtedly instruct us to draw a conclusion Q from a premise P when P is known to be true and $\mathbf{CRs}(Q, P)$ is very large. See Section 24 concerning this rule.

[34] See, to mention only a recent study, J. P. Day, *Inductive Probability.*

[35] See, for example, D. Davidson and P. Suppes, *Decision Making, An Experimental Approach,* and the references supplied therein.

[36] See Carnap, *The Continuum of Inductive Methods,* whose results may be reinterpreted along the lines suggested in the text.

[37] So does Carnap among others, as I mentioned in Note 6 of this chapter. See *Logical Foundations of Probability,* pp. 295–296, on this point.

[38] From now on I occasionally write '$\mathbf{Pi}(P, Q)$' where I once had '$\mathbf{Pi}^N(P, Q)$' or '$\mathbf{Pi}^\infty(P, Q)$.'

[39] D18.1(a1), D18.1(a2′), D18.1(a3), D18.1(a4), D18.1(a5), and D18.1(a6′) are the very requirements which von Wright places upon $\mathbf{Pi}(P, Q)$ in *The Logical Problem of Induction,* pp. 92–93. For a third alternative, under which $\mathbf{Pi}(P, Q)$ comes to equal 0 when Q is logically false in L, see "On logically false evidence statements."

[40] The point is discussed more fully in "On So-called Degrees of Confirmation."

[41] Proof of T22.1 is left to the reader as an exercise.

[42] It pays under alternative (2) to require $Q \,\&\, Q'$ and hence Q not to be logically false in L^N. Otherwise D18.1(a6) would have to read: *If* $Q \,\&\, Q'$ *is not logically false in* L^N, *then*

$$\mathbf{Pi}^N(\sim P, Q) = 1 - \mathbf{Pi}^N(P, Q).$$

[43] See Ramsey, *The Foundations of Mathematics,* pp. 156–211, and de Finetti, "La prévision: ses lois logiques, ses sources subjectives."

[44] On this matter of fairness or coherence, see J. G. Kemeny, "Fair bets and inductive probabilities," R. S. Lehman, "On confirmation and rational betting," and A. Shimony, "Coherence and the axioms of confirmation."

[45] See Davidson and Suppes, *loc. cit.,* pp. 30–40, and Suppes, "Some Open Problems in the Foundations of Subjective Probability."

[46] See Savage, *The Foundations of Statistics,* Chapters 1–4. For still another treatment of personal—or subjective—probabilities, see Davidson and Suppes, "A Finitistic Axiomatization of Subjective Probability and Utility."

[47] For a study of inequalities of the sort $\mathbf{Pi}_0^N(P) \leq \mathbf{Pi}_0^N(Q)$ and, more generally, $\mathbf{Pi}(P, Q) \leq \mathbf{Pi}(P', Q')$, see Carnap, *Logical Foundations of Probability,* chapter VII, and B. O. Koopman, "The axioms and algebra of intuitive probability."

[48] Professor Carnap, after whose suggestions the proposal has been worded, would further require that $\mathbf{Pi}(Q, W = W)$ equal $\mathbf{Pi}(\sim Q, W = W)$. Note with

respect to the third condition in the text that if **Pi**(Q, P & R) were not equal to **Pi**(Q, P), then R would commonly be rated either favorably or unfavorably relevant to Q given P.

[49] In symbols, $\vdash (\cdots(Q_1 \vee Q_2) \vee \cdots) \vee Q_n$ and $\vdash \sim (Q_j \;\&\; Q_k)$ for any two j and k ($j \neq k$) from 1 to n.

[50] See *loc. cit.*, p. 495, and *Notes on Probability and Induction*, p. (11). Carnap's own version of the theorem is slightly stronger; it also holds in L^∞, as Carnap shows, when $\mathbf{Pi}^N$ meets an extra requirement which for lack of space I cannot reproduce here.

[51] In the terminology of Section 10, the sample concerned is an (a)-sample. No other kind of sample can come up for mention in L.

[52] 'A,' 'B,' 'B_1,' and 'B_2' range here as in Chapter 1 over the set abstracts of L.

[53] Proof of T24.3 will be supplied in *Truth and Estimated Truth*.

[54] For an elementary exposition of their results, see Neyman, *First Course in Probability and Statistics*, Chapter 5.

[55] The point will be fully documented in *Truth and Estimated Truth*.

LIST OF SYMBOLS

Page 1: L^∞, L^N, and L
Page 2: $\sim$, $\supset$, $\forall$, $=$, the comma, (,), w, x, y, and z
Page 4: P, Q, R, S, W, X, Y, Z, and G
Page 8: &, v, $\equiv$, $\exists$, and D
Page 11: $\vee^n_{L^\infty}$, $\vee^n_{L^N}$, and $\vee^n_L$
Page 14: $\mathbf{Tv}^\infty$, $\mathbf{Tv}^N$, **Tv**, and **Asst**
Page 17: $\vdash^\infty$, $\vdash^N$, and $\vdash$
Page 21: $\in$ and ˆ
Page 23: A, B, C, D, $=$, $\subset$, and $\emptyset$
Page 24: $-$, $\cap$, and $\cup$
Page 25: P, $\vee$, and $\wedge$
Page 26: {, }, <, and >
Page 27: S
Page 28: Rf
Page 41: **w**
Page 42: **Ps**
Page 68: $\mathbf{Ps}^N$ and $\mathbf{w}^N$
Page 69: $\mathbf{Ps}^\infty$
Page 86: F, V_P, and V_I
Page 91: $\mathbf{CRs}^\infty$, $\mathbf{CRs}^N$, and **CRs**
Page 98: $\mathbf{Pi}^\infty$, $\mathbf{Pi}^N$, and **Pi**
Page 104: $\mathbf{Pi}_0^N$
Page 108: *SD* and **Hv**

BIBLIOGRAPHICAL REFERENCES

The following list includes only such publications as have been mentioned or alluded to in the course of the book. For fuller bibliographical data on probability, see Carnap, R., *Logical Foundations of Probability*, pp. 583–598, Keynes, J. M., *A Treatise on Probability*, pp. 431–458, and von Wright, G. H., *The Logical Problem of Induction*, pp. 227–249.

Ajdukiewicz, K., "La notion de rationalité des méthodes d'inférence faillibles," *Logique et Analyse*, no. 5 (1959), 3–18.

Bolzano, B., *Wissenschaftslehre*. Leipzig: Felix Meimer, 1915 (first published Sulzbach, 1837).

Carnap, R., *Logical Foundations of Probability*. Chicago: U. of Chicago Press, 1950 (2nd ed. forthcoming).

——— *The Continuum of Inductive Methods*. Chicago: U. of Chicago Press, 1952.

——— *Notes on Probability and Induction*, mimeographed notes, The University of California at Los Angeles, 1955.

——— *An Axiom System of Inductive Logic*, mimeographed notes, The University of California at Los Angeles, 1959–1961.

——— *Induktive Logik und Wahrscheinlichkeit*, in collaboration with W. Stegmüller. Vienna: Springer Verlag, 1959.

——— "Replies and Systematic Expositions," in *The Philosophy of Rudolf Carnap*, ed. P. A. Schilpp (forthcoming).

Church, A., *Introduction to Mathematical Logic, Volume I*. Princeton, N. J.: Princeton U. Press, 1956.

Davidson, D. and P. Suppes, *Decision Making, An Experimental Approach*, in collaboration with S. Siegel. Stanford, Calif.: Stanford U. Press, 1957.

——— "A Finitistic Axiomatization of Subjective Probability and Utility," *Econometrica*, XXIV (1956), 264–275.

Day, J. P., *Inductive Probability*. New York: Humanities Press, Inc., 1961.

de Finetti, B., "La prévision: ses lois logiques, ses sources subjectives," *Annales de l'Institut Henri Poincaré*, VII (1937), 1–68.

Feller, W., *An Introduction to Probability Theory and Its Applications, Volume I*, 2nd ed. New York: John Wiley & Sons, Inc., 1957.

Fraenkel, A. A., *Abstract Set Theory*. Amsterdam: North-Holland Publishing Co., 1953.

Goldberg, S., *Probability, An Introduction*. Englewood Cliffs, N. J.: Prentice-Hall, Inc., 1960.

Good, I. J., *Probability and the Weighing of Evidence*. London: Charles Griffen & Co. Ltd., 1950.

Helmer, O. and P. Oppenheim, "A syntactical definition of probability and of degree of confirmation," *The Journal of Symbolic Logic* (1945), X, 25–60.

Hempel, C. G. and P. Oppenheim, "A Definition of "Degree of Confirmation"," *Philosophy of Science*, XII (1945), 98–115.

Hosiasson(-Lindenbaum), J., "On confirmation," *The Journal of Symbolic Logic* V (1940), 133–148.

Jeffreys, H., *Theory of Probability*, 2nd ed. New York: Oxford U. Press, 1948.

Kemeny, J. G., "A logical measure function," *The Journal of Symbolic Logic*, XVIII (1953), 289–308.

——— "Fair bets and inductive probabilities," *ibid.*, XX (1955), 263–273.

Kemeny, J. G., H. Mirkil, J. L. Snell, and G. L. Thompson, *Finite Mathematical Structures*. Englewood Cliffs, N. J.: Prentice-Hall, Inc., 1959.

Keynes, J. M., *A Treatise on Probability*. London: Macmillan & Co., 1921.

Kolmogorov, A. N., *Foundations of the Theory of Probability*. New York: Chelsea Publishing Co., 1950.

Koopman, B. O., "The axioms and algebra of intuitive probability," *Annals of Mathematics*, Series 2, XLI (1940), 269–292.

Kyburg, H. E., Jr., *Probability and the Logic of Rational Belief*. Middletown, Conn.: Wesleyan U. Press, 1961.

Leblanc, H., *An Introduction to Deductive Logic*. New York: John Wiley & Sons, Inc., 1955.

——— "Two Probability Concepts," *The Journal of Philosophy*, LIII (1956), 679–688.

——— "On logically false evidence statements," *The Journal of Symbolic Logic*, XXII (1957), 345–349.

——— "On chances and estimated chances of being true," *Revue Philosophique de Louvain*, LVII (1959), 225–239.

——— "On requirements for conditional probability functions," *The Journal of Symbolic Logic*, XXV (1960), 238–242.

——— "On So-called Degrees of Confirmation," *The British Journal for the Philosophy of Science*, X (1960), 312–315.

——— "On a recent allotment of probabilities to open and closed sentences," *Notre Dame Journal of Formal Logic*, I (1960), 171–175.

——— "A New Interpretation of $\mathsf{c}(h,e)$," *Philosophy and Phenomenological Research*, XXI (1961), 373–376.

——— "The Problem of the Confirmation of Laws," *Philosophical Studies*, XII (1961), 81–84.

——— "Probabilities as Truth-Value Estimates," *Philosophy of Science*, XXVIII (1961), 414–417.

——— "Statistical and Inductive Probabilities," in *Induction: Some Current Issues*, ed. H. E. Kyburg, Jr. and E. Nagel (forthcoming).

——— *Truth and Estimated Truth* (forthcoming).

Lehman, R. S., "On confirmation and rational betting," *The Journal of Symbolic Logic*, XX (1955), 251–262.

Lenz, J. H., "The Frequency Theory of Probability," in *The Structure of*

Scientific Thought, An Introduction to Philosophy of Science, ed. E. H. Madden. Boston: Houghton Mifflin Co., 1960.

Lewis, C. I., *An Analysis of Knowledge and Valuation.* La Salle, Ill.: Open Court Publishing Co., 1946.

Loève, M., *Probability Theory,* 2nd ed. Princeton, N. J.: D. Van Nostrand Co., Inc., 1960.

Menger, K., "Random Variables from the Point of View of a General Theory of Variables," in *Proceedings of the Third Berkeley Symposium on Mathematical Statistics and Probability,* II, 215–229. Berkeley, Calif.: U. of California Press, 1956.

Munroe, M. E., *Theory of Probability.* New York: McGraw-Hill Book Co., Inc., 1951.

Nagel, E., *Principles of the Theory of Probability,* in *International Encyclopedia of Unified Science,* I, no. 6 (1939).

Neyman, J., *First Course in Probability and Statistics.* New York: Holt, Rinehart & Winston, Inc., 1950.

——— *Lectures and Conferences on Mathematical Statistics and Probability,* 2nd ed. Washington: Graduate School, U. S. Department of Agriculture, 1952.

Parzen, E., *Modern Probability Theory and Its Applications.* New York: John Wiley & Sons, Inc., 1960.

Peirce, C. S., *Collected Papers of Charles Sanders Peirce.* Cambridge, Mass.: Harvard U. Press, 1931–1958.

Popper, K. R., *The Logic of Scientific Discovery.* New York: Basic Books, Inc., 1959.

Quine, W. V. O., *Mathematical Logic,* rev. ed. Cambridge, Mass.: Harvard U. Press, 1951.

——— *Methods of Logic,* rev. ed. New York: Holt, Rinehart & Winston, Inc., 1959.

Ramsey, F. P., *The Foundations of Mathematics.* London: Routledge & Kegan Paul Ltd., 1931.

Reichenbach, H., *The Theory of Probability.* Berkeley, Calif.: U. of California Press, 1949.

Russell, B., *Human Knowledge, Its Scope and Limits.* New York: Simon and Schuster, Inc., 1948.

Savage, L. J., *The Foundations of Statistics.* John Wiley & Sons, Inc., 1954.

Shimony, A., "Coherence and the axioms of confirmation," *The Journal of Symbolic Logic,* XX (1955), 1–28.

Suppes, P., *Axiomatic Set Theory.* Princeton, N. J.: D. Van Nostrand Co., Inc., 1960.

——— "Some Open Problems in the Foundations of Subjective Probability," in *Information and Decision Processes,* ed. R. E. Machol (1960).

Tarski, A., "The Semantic Conception of Truth and the Foundations of Semantics," *Philosophy and Phenomenological Research,* IV (1944), 13–47.

von Mises, R., "Grundlagen der Wahrscheinlichkeitsrechnung," *Mathematische Zeitschrift,* V (1919), 52–99.

——— *Probability, Statistics and Truth.* New York: The Macmillan Co., 1939.

von Wright, G. H., "Carnap's Theory of Probability," *The Philosophical Review,* LX (1951), 362–374.

——— *A Treatise on Induction and Probability.* New York: Harcourt, Brace and Co., 1951.

——— *The Logical Problem of Induction,* 2nd rev. ed. New York: The Macmillan Co., 1957.

Waismann, F., "Logische Analyse des Wahrscheinlichkeitsbegriffs," *Erkenntnis,* I (1930–1931), 228–248.

Wilder, R. L., *Introduction to the Foundations of Mathematics.* New York: John Wiley & Sons, Inc., 1952.

Williams, D. C., *The Ground of Induction.* Cambridge, Mass.: Harvard U. Press, 1947.

Wittgenstein, L., *Tractatus Logico-Philosophicus.* London: Routledge & Kegan Paul Ltd., 1961 (first published in London, 1922).

INDEX OF AUTHORS

Ajdukiewicz, K., 95, 96
Bolzano, B., 95
Carnap, R., 29, 63, 65–66, 97, 98, 106, 107, 111–112, 120, 131–132, 134, 135, 136, 137, 138, 139
Church, A., 29
Davidson, D., 138
Day, J. P., 138
de Finetti, B., 97, 127, 129, 138
Feller, W., 39, 51, 61, 63, 64, 65, 66
Fraenkel, A. A., 29, 30, 62
Gödel, K., 16, 31
Goldberg, S., 62
Good, I. J., 134
Helmer, O., 136, 137
Hempel, C. G., 136, 137
Hermes, H., 63
Hosiasson, J., 134
Jeffrey, R., 94
Jeffreys, H., 134
Kanger, S., 31, 94
Kemeny, J. G., 62, 64, 68, 86-87, 93, 95, 110, 136, 138
Keynes, J. M., 112, 127, 134
Kolmogorov, A. N., 39, 51, 61, 62, 63, 65, 66
Koopman, B. O., 138
Kyburg, H. E. Jr., 63
Lehman, R. S., 138
Lenz, J. H., 96
Lewis, C. I., 137
Loève, M., 39, 51, 61, 62, 63, 65, 66
Menger, K., 64
Mirkil, H., 62, 64, 93, 95
Munroe, M. E., 62
Nagel, E., 40, 62, 63, 64
Neyman, J., 32, 38–39, 40, 51, 52, 60. 61, 62, 63, 64, 134, 139
Oppenheim, P., 136, 137
Pap, A., 96
Parzen, E., 39, 51, 61, 63, 65, 66
Pearson, E. S., 134
Peirce, C. S., 95, 112
Popper, K. R., 135, 136
Quine, W. V., 29
Ramsey, F. P., 97, 127, 129, 138
Reichenbach, H., 37, 62
Russell, B., 62, 95, 96
Savage, L. J., 97, 128–129, 138
Sheffer, I., 137
Shimony, A., 138
Snell, J. L., 62, 64, 93, 95
Sorenson, R. T., 63
Spencer, N., 31
Suppes, P., 29, 32, 128, 138
Tarski, A., 30, 31
Thompson, G. L., 62, 64, 93, 95
Ullian, J., 95
von Mises, R., 37, 62, 65
von Wright, G. H., 134, 135, 136, 137, 138
Waismann, F., 135–136
Wilder, R. L., 32
Williams, D. C., 95
Wittgenstein, L., 112

INDEX OF MATTERS

A, 38–39
Allotments of probabilities and weights:
  equiprobable, *see* relative frequency
  finite-frequency, *see* relative frequency
  number of, 35–36, 11, 120
  regular, 106–107, 132–134
  relative frequency, 34–37, 48–51, 57–61, 72, 74–85, 87–88, 90–93, 111–112, 116–120
  strongly regular, 106, 111, 131–134
  symmetrical, 111–112, 131–134
  weakly regular, 106, 111
Betting, 121, 126–129
  habits, 121, 129
  quotients, 121, 126–128, 137
Binomial distribution, 54–57
Cartesian product, 62
Coefficient of statistical reliability, 90–93
  estimation of, 93, 129–130
Comma, 2–3
Conclusion, 20–21
Conditionals, 6
Connectives, 2–3, 8
Deducibility, 20–21
Defined signs, 8–9, 21–29, 31
Degrees of confirmation, 123–124, 137
Estimation:
  of coefficient of statistical reliability, 93, 129–130
  of values of random functions, 116
  of statistical probabilities, 57–61, 78, 112–121
  of truth-values, 112–126
Evidence sentences, 122–126
Experiments, 39–40
  probabilities allotted to sets of outcomes of, 39–45, 49–51
  probability sets generated by, 49–51
  weights allotted to outcomes of, 39–45, 49–51, 59–61
Expressions, 5
Fair coin, 44–45, 59–60
Falsehood, 14
  logical, 18, 107, 122–123, 315
Functions, 51
  numerical, 51
  probability, *see* Inductive probability functions
  random, *see* Random functions
Functors, 27–29, 86–87
Holding-values, 109–110
Hypergeometric distribution, 54–57, 131–134
Hypergeometric theorem, 54–57, 131–134
Identities, 6
Identity sign, 2–3
Indefiniteness as to subject matter, 39–40, 47, 85
Individual-constant-free mate, 88–92
Individual constants, 2–4, 49
  alphabetical order of, 2
  designations of, 9–13, 16–18
  free (occurrences of), 5–7
  ordering of sequences of, 69–71
  weights allotted to (sequences of), 50–51, 68–76, 111, 120, 130–131
Individual signs, 2
Individual variables, 2–3
  alphabetical order of, 2
  assignments of individuals to, 14–15, 68, 86–87
  bound (occurrences of), 5–7
  free (occurrences of), 5–8
  range of values of, 9
Individuals, 9–12, 48–49
  assigned to individual variables, 14–15, 68, 86–87
  indistinguishable, 49–51
  sequences of, 9–12, 26–27
  weights allotted to (sequences of), 86–87
Inductive inequalities, 128–129
Inductive probabilities:
  absolute, 104–106, 109–112, 125–126
  allotted to sentences, 98–112
  as betting quotients, 121, 126–128
  as degrees of confirmation, 123–124, 137

Inductive probabilities (*cont.*):
as estimates of statistical probabilities, 58, 78, 112–121
as estimates of truth-values, 112–126
conditional, 104–107, 110–112, 125–126
contrasted with personal probabilities, 126–129
contrasted with statistical probabilities and weights, 40, 47, 59–61
inferential uses of, 129–134
interpretation of, 112–127
requirements satisfied by, 98–99, 104–106, 110–115, 120, 122–129
Inductive probability functions:
number of, 111, 120
regular, 106–107, 132–134
relative frequency, 111–112
strongly regular, 106, 111, 131–134
symmetrical, 111–112, 131–134
weakly regular, 106, 111
Inferences:
deductive, 20–21, 89–90
inductive, 41, 89–93, 120–121, 129–134
Instances of sentences, 5–8, 77–78, 116–120
Language L^∞, 1–4, 8, 9–12, 15, 17–18, 25, 28, 49, 69–80, 88–89, 91–94, 98–100, 103–104, 112, 122–123, 125, 132
sublanguages of, 1–4, 8, 10–11, 15, 17–18, 25, 28, 68–80, 87–94, 97–134
Languages *L*:
grammar of, 4–8
inductive probabilities allotted to sentences of, 97–134
interpretation of, 9–21
metalanguage *ML* of, 27, 38
set theory grafted onto, 21–29
statistical probabilities allotted to sentences of, 67–93
vocabulary of, 2–4, 8–9, 21–29
Logical equivalence, 20, 107
Logical falsehood, 18, 107, 122–123, 135
Logical implication, 20–21, 107
Logical possibilities, 86–87, 110, 128–129
weights allotted to, 86
Logical truth, 15–21, 103, 107, 123–126
Lotteries, 128–129
Mathematical induction, 80
Measurements on sentences and sets, 33, 60–61, 67, 78
Metalinguistic variables, 4–5, 23, 27, 61, 109, 139
Negations, 6
Odds, 126–128
Parentheses, 2–3, 6
Populations, 53–57, 132–134
cells of, 54–57, 132–133
partitions of, 54–57, 132–133
samples from, *see* Samples
Predicates, 2–4
extensions of, 9–12, 16–17, 31
n-place, 3
Premises, 20–21
Primitive signs, 2–4, 27–28, 38, 49
Probabilities:
absolute, 34–38, 47–48, 70, 78, 104–106, 109–112, 125–126
comparative, 129
conditional, 35–37, 47–48, 70, 78, 104–107, 110–112, 125–126
inductive, *see* Inductive probabilities
personal, 126–129
statistical, *see* Statistical probabilities
subjective, *see* personal
Probability functions, *see* Inductive probability functions
Probability pairs of sentences, 122–126
Probability sets:
as sets generally, 34–40, 48–51
as sets of outcomes of experiments, 39–40
denumerably infinite (and serially ordered), 36–38, 70
finite, 34–36, 38, 48, 68
generated by experiments, 49–51
indistinguishable members of, 49–51
infinite, 36–38, 48–49
non-denumerably infinite, 37
probabilities allotted to sets of (sequences of) members of, 34–38
probabilities allotted to subsets of, 34–38
unspecified members of, 38–41
weights allotted to (sequences of) members of, 34–38, 49–51, 120
Proportions, 28–29
Quantifier letters, 2–3, 8–9
Quantifiers, 7
Random drawing, 56–58, 60, 132–134
Random functions (or variables), 51–53
estimation of values of, 116
probabilities and weights allotted to, 52
Random occurrence, 37, 65
Reference sets, 35–36, 47–48
Relative frequencies, 28–29
Relative frequency allotment, *see* Allotments of probabilities and weights
Samples:
(*a*)-samples, 53–57, 132–134
(*b*1)-samples, 53–57
(*b*2)-samples, 53–57

Samples (*cont.*):
 drawing of, 53–55
 random drawing of, 56–58, 60, 132–134
 weights allotted to drawings of, 54–57, 60–61
Satisfaction, 14–15
Sentences, 5–6
 closed, 5–8
 elementary, 108
 false, 14
 inductive probabilities allotted to, 98–112
 instances of, 5–8, 77–78, 116–120
 logically false, 18, 107, 122–123, 135
 logically true, 15–21, 107, 123–126
 open, 5–8
 quantifier-and-identity-free, 106–107
 statistical probabilities allotted to, 38–39, 68–76, 85–89
 true, 12–17
 universal, 6
 valid, 16–20
Set(s):
 abstracts, 21–23, 38, 68–70
 cells of, 25
 complements of, 24–25
 defining conditions of, 21–23, 38
 denumerably infinite, 28–29
 given by enumeration, 26
 identity of, 23
 inclusion of, 23
 intersections of, 24–25
 membership in, 21–23, 38–41
 non-overlapping of, 23
 null, 25–26, 63, 68
 partitions of, 25
 probabilities allotted to, 34–38
 serially ordered, 28–29
 sizes of, 24–28
 subsets of, 23–24
 unions of, 24–25
 unit, 26
 universal, 25–26, 38, 68
 unspecified members of, 38–41
Single-case issue, 46–47, 61, 89
State-descriptions, 108–112, 128–129
 weights allotted to, 109–112, 120, 134
Statistical probabilities:
 absolute, 34–38, 47–48, 70, 78
 allotted to random functions, 52
 allotted to sentences, 38–39, 68–76, 85–89
 allotted to sets, 34–38
 allotted to sets of outcomes of experiments, 39–45, 49–51
Statistical probabilities (*cont.*):
 allotted to sets of (sequences of) members of probability sets, 34–38
 allotted to subsets of probability sets, 34–38
 as truth-values, 76–78
 conditional, 35–37, 47–48, 70, 78
 contrasted with inductive probabilities, 40, 47, 61
 estimation of, 57–61, 78, 112–121
 inferential uses of, 41, 89–93
 interpretation of, 38–41, 46–47, 76–78
 requirements satisfied by, 34, 36, 78–85
Stochastic independence, 45, 61
Truth, 12–17
 logical, 15–21, 107, 123–126
Truth-values, 14, 76–78
 average, 77–78
 estimation of, 112–126
Uncertainty as to truth-value, 40, 112
Universes of discourse, 3–4, 9–13, 18
 denumerably infinite, 4
 finite, 4
 non-denumerably infinite, 29
 sizes of, 25–26, 29
Validity, 16–20
Weights:
 allotted to arguments of random functions, 52
 allotted to drawings of samples, 54–57, 60–61
 allotted to logical possibilities, 86
 allotted to outcomes of experiments, 39–45, 49–51, 59–61
 allotted to (sequences of) individual constants, 50–51, 68–76, 111, 120, 130–131
 allotted to (sequences of) individuals, 86–87
 allotted to (sequences of) members of probability sets, 34–38, 49–51, 120
 allotted to state-descriptions, 109–112, 120, 134
 contrasted with estimates of statistical probabilities, 59–61
 converted into statistical probabilities, 72–73
 equal versus unequal, 44–45, 48–51, 54–57, 59–61, 111
 interpretation of, 38–41, 60–61, 93–94
 requirements satisfied by, 34, 71, 110–111

Astronomy

BURNHAM'S CELESTIAL HANDBOOK, Robert Burnham, Jr. Thorough guide to the stars beyond our solar system. Exhaustive treatment. Alphabetical by constellation: Andromeda to Cetus in Vol. 1; Chamaeleon to Orion in Vol. 2; and Pavo to Vulpecula in Vol. 3. Hundreds of illustrations. Index in Vol. 3. 2,000pp. 6⅛ x 9¼.
Vol. I: 0-486-23567-X
Vol. II: 0-486-23568-8
Vol. III: 0-486-23673-0

EXPLORING THE MOON THROUGH BINOCULARS AND SMALL TELESCOPES, Ernest H. Cherrington, Jr. Informative, profusely illustrated guide to locating and identifying craters, rills, seas, mountains, other lunar features. Newly revised and updated with special section of new photos. Over 100 photos and diagrams. 240pp. 8¼ x 11. 0-486-24491-1

THE EXTRATERRESTRIAL LIFE DEBATE, 1750–1900, Michael J. Crowe. First detailed, scholarly study in English of the many ideas that developed from 1750 to 1900 regarding the existence of intelligent extraterrestrial life. Examines ideas of Kant, Herschel, Voltaire, Percival Lowell, many other scientists and thinkers. 16 illustrations. 704pp. 5⅜ x 8½. 0-486-40675-X

THEORIES OF THE WORLD FROM ANTIQUITY TO THE COPERNICAN REVOLUTION, Michael J. Crowe. Newly revised edition of an accessible, enlightening book recreates the change from an earth-centered to a sun-centered conception of the solar system. 242pp. 5⅜ x 8½. 0-486-41444-2

A HISTORY OF ASTRONOMY, A. Pannekoek. Well-balanced, carefully reasoned study covers such topics as Ptolemaic theory, work of Copernicus, Kepler, Newton, Eddington's work on stars, much more. Illustrated. References. 521pp. 5⅜ x 8½.
0-486-65994-1

A COMPLETE MANUAL OF AMATEUR ASTRONOMY: TOOLS AND TECHNIQUES FOR ASTRONOMICAL OBSERVATIONS, P. Clay Sherrod with Thomas L. Koed. Concise, highly readable book discusses: selecting, setting up and maintaining a telescope; amateur studies of the sun; lunar topography and occultations; observations of Mars, Jupiter, Saturn, the minor planets and the stars; an introduction to photoelectric photometry; more. 1981 ed. 124 figures. 25 halftones. 37 tables. 335pp. 6½ x 9¼. 0-486-40675-X

AMATEUR ASTRONOMER'S HANDBOOK, J. B. Sidgwick. Timeless, comprehensive coverage of telescopes, mirrors, lenses, mountings, telescope drives, micrometers, spectroscopes, more. 189 illustrations. 576pp. 5⅝ x 8¼. (Available in U.S. only.)
0-486-24034-7

STARS AND RELATIVITY, Ya. B. Zel'dovich and I. D. Novikov. Vol. 1 of *Relativistic Astrophysics* by famed Russian scientists. General relativity, properties of matter under astrophysical conditions, stars, and stellar systems. Deep physical insights, clear presentation. 1971 edition. References. 544pp. 5⅝ x 8¼. 0-486-69424-0

Chemistry

THE SCEPTICAL CHYMIST: THE CLASSIC 1661 TEXT, Robert Boyle. Boyle defines the term "element," asserting that all natural phenomena can be explained by the motion and organization of primary particles. 1911 ed. viii+232pp. 5⅜ x 8½. 0-486-42825-7

RADIOACTIVE SUBSTANCES, Marie Curie. Here is the celebrated scientist's doctoral thesis, the prelude to her receipt of the 1903 Nobel Prize. Curie discusses establishing atomic character of radioactivity found in compounds of uranium and thorium; extraction from pitchblende of polonium and radium; isolation of pure radium chloride; determination of atomic weight of radium; plus electric, photographic, luminous, heat, color effects of radioactivity. ii+94pp. 5⅜ x 8½. 0-486-42550-9

CHEMICAL MAGIC, Leonard A. Ford. Second Edition, Revised by E. Winston Grundmeier. Over 100 unusual stunts demonstrating cold fire, dust explosions, much more. Text explains scientific principles and stresses safety precautions. 128pp. 5⅜ x 8½. 0-486-67628-5

THE DEVELOPMENT OF MODERN CHEMISTRY, Aaron J. Ihde. Authoritative history of chemistry from ancient Greek theory to 20th-century innovation. Covers major chemists and their discoveries. 209 illustrations. 14 tables. Bibliographies. Indices. Appendices. 851pp. 5⅜ x 8½. 0-486-64235-6

CATALYSIS IN CHEMISTRY AND ENZYMOLOGY, William P. Jencks. Exceptionally clear coverage of mechanisms for catalysis, forces in aqueous solution, carbonyl- and acyl-group reactions, practical kinetics, more. 864pp. 5⅜ x 8½. 0-486-65460-5

ELEMENTS OF CHEMISTRY, Antoine Lavoisier. Monumental classic by founder of modern chemistry in remarkable reprint of rare 1790 Kerr translation. A must for every student of chemistry or the history of science. 539pp. 5⅜ x 8½. 0-486-64624-6

THE HISTORICAL BACKGROUND OF CHEMISTRY, Henry M. Leicester. Evolution of ideas, not individual biography. Concentrates on formulation of a coherent set of chemical laws. 260pp. 5⅜ x 8½. 0-486-61053-5

A SHORT HISTORY OF CHEMISTRY, J. R. Partington. Classic exposition explores origins of chemistry, alchemy, early medical chemistry, nature of atmosphere, theory of valency, laws and structure of atomic theory, much more. 428pp. 5⅜ x 8½. (Available in U.S. only.) 0-486-65977-1

GENERAL CHEMISTRY, Linus Pauling. Revised 3rd edition of classic first-year text by Nobel laureate. Atomic and molecular structure, quantum mechanics, statistical mechanics, thermodynamics correlated with descriptive chemistry. Problems. 992pp. 5⅜ x 8½. 0-486-65622-5

FROM ALCHEMY TO CHEMISTRY, John Read. Broad, humanistic treatment focuses on great figures of chemistry and ideas that revolutionized the science. 50 illustrations. 240pp. 5⅜ x 8½. 0-486-28690-8

Engineering

DE RE METALLICA, Georgius Agricola. The famous Hoover translation of greatest treatise on technological chemistry, engineering, geology, mining of early modern times (1556). All 289 original woodcuts. 638pp. 6¾ x 11. 0-486-60006-8

FUNDAMENTALS OF ASTRODYNAMICS, Roger Bate et al. Modern approach developed by U.S. Air Force Academy. Designed as a first course. Problems, exercises. Numerous illustrations. 455pp. 5⅜ x 8½. 0-486-60061-0

DYNAMICS OF FLUIDS IN POROUS MEDIA, Jacob Bear. For advanced students of ground water hydrology, soil mechanics and physics, drainage and irrigation engineering and more. 335 illustrations. Exercises, with answers. 784pp. 6⅛ x 9¼. 0-486-65675-6

THEORY OF VISCOELASTICITY (Second Edition), Richard M. Christensen. Complete consistent description of the linear theory of the viscoelastic behavior of materials. Problem-solving techniques discussed. 1982 edition. 29 figures. xiv+364pp. 6⅛ x 9¼. 0-486-42880-X

MECHANICS, J. P. Den Hartog. A classic introductory text or refresher. Hundreds of applications and design problems illuminate fundamentals of trusses, loaded beams and cables, etc. 334 answered problems. 462pp. 5⅜ x 8½. 0-486-60754-2

MECHANICAL VIBRATIONS, J. P. Den Hartog. Classic textbook offers lucid explanations and illustrative models, applying theories of vibrations to a variety of practical industrial engineering problems. Numerous figures. 233 problems, solutions. Appendix. Index. Preface. 436pp. 5⅜ x 8½. 0-486-64785-4

STRENGTH OF MATERIALS, J. P. Den Hartog. Full, clear treatment of basic material (tension, torsion, bending, etc.) plus advanced material on engineering methods, applications. 350 answered problems. 323pp. 5⅜ x 8½. 0-486-60755-0

A HISTORY OF MECHANICS, René Dugas. Monumental study of mechanical principles from antiquity to quantum mechanics. Contributions of ancient Greeks, Galileo, Leonardo, Kepler, Lagrange, many others. 671pp. 5⅜ x 8½. 0-486-65632-2

STABILITY THEORY AND ITS APPLICATIONS TO STRUCTURAL MECHANICS, Clive L. Dym. Self-contained text focuses on Koiter postbuckling analyses, with mathematical notions of stability of motion. Basing minimum energy principles for static stability upon dynamic concepts of stability of motion, it develops asymptotic buckling and postbuckling analyses from potential energy considerations, with applications to columns, plates, and arches. 1974 ed. 208pp. 5⅜ x 8½. 0-486-42541-X

METAL FATIGUE, N. E. Frost, K. J. Marsh, and L. P. Pook. Definitive, clearly written, and well-illustrated volume addresses all aspects of the subject, from the historical development of understanding metal fatigue to vital concepts of the cyclic stress that causes a crack to grow. Includes 7 appendixes. 544pp. 5⅜ x 8½. 0-486-40927-9

ROCKETS, Robert Goddard. Two of the most significant publications in the history of rocketry and jet propulsion: "A Method of Reaching Extreme Altitudes" (1919) and "Liquid Propellant Rocket Development" (1936). 128pp. 5⅜ x 8½. 0-486-42537-1

STATISTICAL MECHANICS: PRINCIPLES AND APPLICATIONS, Terrell L. Hill. Standard text covers fundamentals of statistical mechanics, applications to fluctuation theory, imperfect gases, distribution functions, more. 448pp. 5⅜ x 8½. 0-486-65390-0

ENGINEERING AND TECHNOLOGY 1650–1750: ILLUSTRATIONS AND TEXTS FROM ORIGINAL SOURCES, Martin Jensen. Highly readable text with more than 200 contemporary drawings and detailed engravings of engineering projects dealing with surveying, leveling, materials, hand tools, lifting equipment, transport and erection, piling, bailing, water supply, hydraulic engineering, and more. Among the specific projects outlined-transporting a 50-ton stone to the Louvre, erecting an obelisk, building timber locks, and dredging canals. 207pp. 8⅜ x 11¼. 0-486-42232-1

THE VARIATIONAL PRINCIPLES OF MECHANICS, Cornelius Lanczos. Graduate level coverage of calculus of variations, equations of motion, relativistic mechanics, more. First inexpensive paperbound edition of classic treatise. Index. Bibliography. 418pp. 5⅜ x 8½. 0-486-65067-7

PROTECTION OF ELECTRONIC CIRCUITS FROM OVERVOLTAGES, Ronald B. Standler. Five-part treatment presents practical rules and strategies for circuits designed to protect electronic systems from damage by transient overvoltages. 1989 ed. xxiv+434pp. 6⅛ x 9¼. 0-486-42552-5

ROTARY WING AERODYNAMICS, W. Z. Stepniewski. Clear, concise text covers aerodynamic phenomena of the rotor and offers guidelines for helicopter performance evaluation. Originally prepared for NASA. 537 figures. 640pp. 6⅛ x 9¼. 0-486-64647-5

INTRODUCTION TO SPACE DYNAMICS, William Tyrrell Thomson. Comprehensive, classic introduction to space-flight engineering for advanced undergraduate and graduate students. Includes vector algebra, kinematics, transformation of coordinates. Bibliography. Index. 352pp. 5⅜ x 8½. 0-486-65113-4

HISTORY OF STRENGTH OF MATERIALS, Stephen P. Timoshenko. Excellent historical survey of the strength of materials with many references to the theories of elasticity and structure. 245 figures. 452pp. 5⅜ x 8½. 0-486-61187-6

ANALYTICAL FRACTURE MECHANICS, David J. Unger. Self-contained text supplements standard fracture mechanics texts by focusing on analytical methods for determining crack-tip stress and strain fields. 336pp. 6⅛ x 9¼. 0-486-41737-9

STATISTICAL MECHANICS OF ELASTICITY, J. H. Weiner. Advanced, self-contained treatment illustrates general principles and elastic behavior of solids. Part 1, based on classical mechanics, studies thermoelastic behavior of crystalline and polymeric solids. Part 2, based on quantum mechanics, focuses on interatomic force laws, behavior of solids, and thermally activated processes. For students of physics and chemistry and for polymer physicists. 1983 ed. 96 figures. 496pp. 5⅜ x 8½. 0-486-42260-7

Mathematics

FUNCTIONAL ANALYSIS (Second Corrected Edition), George Bachman and Lawrence Narici. Excellent treatment of subject geared toward students with background in linear algebra, advanced calculus, physics and engineering. Text covers introduction to inner-product spaces, normed, metric spaces, and topological spaces; complete orthonormal sets, the Hahn-Banach Theorem and its consequences, and many other related subjects. 1966 ed. 544pp. 6⅛ x 9¼. 0-486-40251-7

ASYMPTOTIC EXPANSIONS OF INTEGRALS, Norman Bleistein & Richard A. Handelsman. Best introduction to important field with applications in a variety of scientific disciplines. New preface. Problems. Diagrams. Tables. Bibliography. Index. 448pp. 5⅜ x 8½. 0-486-65082-0

VECTOR AND TENSOR ANALYSIS WITH APPLICATIONS, A. I. Borisenko and I. E. Tarapov. Concise introduction. Worked-out problems, solutions, exercises. 257pp. 5⅝ x 8¼. 0-486-63833-2

AN INTRODUCTION TO ORDINARY DIFFERENTIAL EQUATIONS, Earl A. Coddington. A thorough and systematic first course in elementary differential equations for undergraduates in mathematics and science, with many exercises and problems (with answers). Index. 304pp. 5⅜ x 8½. 0-486-65942-9

FOURIER SERIES AND ORTHOGONAL FUNCTIONS, Harry F. Davis. An incisive text combining theory and practical example to introduce Fourier series, orthogonal functions and applications of the Fourier method to boundary-value problems. 570 exercises. Answers and notes. 416pp. 5⅜ x 8½. 0-486-65973-9

COMPUTABILITY AND UNSOLVABILITY, Martin Davis. Classic graduate-level introduction to theory of computability, usually referred to as theory of recurrent functions. New preface and appendix. 288pp. 5⅜ x 8½. 0-486-61471-9

ASYMPTOTIC METHODS IN ANALYSIS, N. G. de Bruijn. An inexpensive, comprehensive guide to asymptotic methods–the pioneering work that teaches by explaining worked examples in detail. Index. 224pp. 5⅜ x 8½ 0-486-64221-6

APPLIED COMPLEX VARIABLES, John W. Dettman. Step-by-step coverage of fundamentals of analytic function theory–plus lucid exposition of five important applications: Potential Theory; Ordinary Differential Equations; Fourier Transforms; Laplace Transforms; Asymptotic Expansions. 66 figures. Exercises at chapter ends. 512pp. 5⅜ x 8½. 0-486-64670-X

INTRODUCTION TO LINEAR ALGEBRA AND DIFFERENTIAL EQUATIONS, John W. Dettman. Excellent text covers complex numbers, determinants, orthonormal bases, Laplace transforms, much more. Exercises with solutions. Undergraduate level. 416pp. 5⅜ x 8½. 0-486-65191-6

RIEMANN'S ZETA FUNCTION, H. M. Edwards. Superb, high-level study of landmark 1859 publication entitled "On the Number of Primes Less Than a Given Magnitude" traces developments in mathematical theory that it inspired. xiv+315pp. 5⅜ x 8½. 0-486-41740-9

INTRODUCTORY REAL ANALYSIS, A.N. Kolmogorov, S. V. Fomin. Translated by Richard A. Silverman. Self-contained, evenly paced introduction to real and functional analysis. Some 350 problems. 403pp. 5⅜ x 8½. 0-486-61226-0

APPLIED ANALYSIS, Cornelius Lanczos. Classic work on analysis and design of finite processes for approximating solution of analytical problems. Algebraic equations, matrices, harmonic analysis, quadrature methods, much more. 559pp. 5⅜ x 8½. 0-486-65656-X

AN INTRODUCTION TO ALGEBRAIC STRUCTURES, Joseph Landin. Superb self-contained text covers "abstract algebra": sets and numbers, theory of groups, theory of rings, much more. Numerous well-chosen examples, exercises. 247pp. 5⅜ x 8½. 0-486-65940-2

QUALITATIVE THEORY OF DIFFERENTIAL EQUATIONS, V. V. Nemytskii and V.V. Stepanov. Classic graduate-level text by two prominent Soviet mathematicians covers classical differential equations as well as topological dynamics and ergodic theory. Bibliographies. 523pp. 5⅜ x 8½. 0-486-65954-2

THEORY OF MATRICES, Sam Perlis. Outstanding text covering rank, nonsingularity and inverses in connection with the development of canonical matrices under the relation of equivalence, and without the intervention of determinants. Includes exercises. 237pp. 5⅜ x 8½. 0-486-66810-X

INTRODUCTION TO ANALYSIS, Maxwell Rosenlicht. Unusually clear, accessible coverage of set theory, real number system, metric spaces, continuous functions, Riemann integration, multiple integrals, more. Wide range of problems. Undergraduate level. Bibliography. 254pp. 5⅜ x 8½. 0-486-65038-3

MODERN NONLINEAR EQUATIONS, Thomas L. Saaty. Emphasizes practical solution of problems; covers seven types of equations. ". . . a welcome contribution to the existing literature...."–*Math Reviews*. 490pp. 5⅜ x 8½. 0-486-64232-1

MATRICES AND LINEAR ALGEBRA, Hans Schneider and George Phillip Barker. Basic textbook covers theory of matrices and its applications to systems of linear equations and related topics such as determinants, eigenvalues and differential equations. Numerous exercises. 432pp. 5⅜ x 8½. 0-486-66014-1

LINEAR ALGEBRA, Georgi E. Shilov. Determinants, linear spaces, matrix algebras, similar topics. For advanced undergraduates, graduates. Silverman translation. 387pp. 5⅜ x 8½. 0-486-63518-X

ELEMENTS OF REAL ANALYSIS, David A. Sprecher. Classic text covers fundamental concepts, real number system, point sets, functions of a real variable, Fourier series, much more. Over 500 exercises. 352pp. 5⅜ x 8½. 0-486-65385-4

SET THEORY AND LOGIC, Robert R. Stoll. Lucid introduction to unified theory of mathematical concepts. Set theory and logic seen as tools for conceptual understanding of real number system. 496pp. 5⅜ x 8¼. 0-486-63829-4

Math–Decision Theory, Statistics, Probability

ELEMENTARY DECISION THEORY, Herman Chernoff and Lincoln E. Moses. Clear introduction to statistics and statistical theory covers data processing, probability and random variables, testing hypotheses, much more. Exercises. 364pp. 5⅜ x 8½. 0-486-65218-1

STATISTICS MANUAL, Edwin L. Crow et al. Comprehensive, practical collection of classical and modern methods prepared by U.S. Naval Ordnance Test Station. Stress on use. Basics of statistics assumed. 288pp. 5⅜ x 8½. 0-486-60599-X

SOME THEORY OF SAMPLING, William Edwards Deming. Analysis of the problems, theory and design of sampling techniques for social scientists, industrial managers and others who find statistics important at work. 61 tables. 90 figures. xvii +602pp. 5⅜ x 8½. 0-486-64684-X

LINEAR PROGRAMMING AND ECONOMIC ANALYSIS, Robert Dorfman, Paul A. Samuelson and Robert M. Solow. First comprehensive treatment of linear programming in standard economic analysis. Game theory, modern welfare economics, Leontief input-output, more. 525pp. 5⅜ x 8½. 0-486-65491-5

PROBABILITY: AN INTRODUCTION, Samuel Goldberg. Excellent basic text covers set theory, probability theory for finite sample spaces, binomial theorem, much more. 360 problems. Bibliographies. 322pp. 5⅜ x 8½. 0-486-65252-1

GAMES AND DECISIONS: INTRODUCTION AND CRITICAL SURVEY, R. Duncan Luce and Howard Raiffa. Superb nontechnical introduction to game theory, primarily applied to social sciences. Utility theory, zero-sum games, n-person games, decision-making, much more. Bibliography. 509pp. 5⅜ x 8½. 0-486-65943-7

INTRODUCTION TO THE THEORY OF GAMES, J. C. C. McKinsey. This comprehensive overview of the mathematical theory of games illustrates applications to situations involving conflicts of interest, including economic, social, political, and military contexts. Appropriate for advanced undergraduate and graduate courses; advanced calculus a prerequisite. 1952 ed. x+372pp. 5⅜ x 8½. 0-486-42811-7

FIFTY CHALLENGING PROBLEMS IN PROBABILITY WITH SOLUTIONS, Frederick Mosteller. Remarkable puzzlers, graded in difficulty, illustrate elementary and advanced aspects of probability. Detailed solutions. 88pp. 5⅜ x 8½. 65355-2

PROBABILITY THEORY: A CONCISE COURSE, Y. A. Rozanov. Highly readable, self-contained introduction covers combination of events, dependent events, Bernoulli trials, etc. 148pp. 5⅜ x 8¼. 0-486-63544-9

STATISTICAL METHOD FROM THE VIEWPOINT OF QUALITY CONTROL, Walter A. Shewhart. Important text explains regulation of variables, uses of statistical control to achieve quality control in industry, agriculture, other areas. 192pp. 5⅜ x 8½. 0-486-65232-7

Math–Geometry and Topology

ELEMENTARY CONCEPTS OF TOPOLOGY, Paul Alexandroff. Elegant, intuitive approach to topology from set-theoretic topology to Betti groups; how concepts of topology are useful in math and physics. 25 figures. 57pp. 5⅜ x 8½. 0-486-60747-X

COMBINATORIAL TOPOLOGY, P. S. Alexandrov. Clearly written, well-organized, three-part text begins by dealing with certain classic problems without using the formal techniques of homology theory and advances to the central concept, the Betti groups. Numerous detailed examples. 654pp. 5⅜ x 8½. 0-486-40179-0

EXPERIMENTS IN TOPOLOGY, Stephen Barr. Classic, lively explanation of one of the byways of mathematics. Klein bottles, Moebius strips, projective planes, map coloring, problem of the Koenigsberg bridges, much more, described with clarity and wit. 43 figures. 210pp. 5⅜ x 8½. 0-486-25933-1

THE GEOMETRY OF RENÉ DESCARTES, René Descartes. The great work founded analytical geometry. Original French text, Descartes's own diagrams, together with definitive Smith-Latham translation. 244pp. 5⅜ x 8½. 0-486-60068-8

EUCLIDEAN GEOMETRY AND TRANSFORMATIONS, Clayton W. Dodge. This introduction to Euclidean geometry emphasizes transformations, particularly isometries and similarities. Suitable for undergraduate courses, it includes numerous examples, many with detailed answers. 1972 ed. viii+296pp. 6⅛ x 9¼. 0-486-43476-1

PRACTICAL CONIC SECTIONS: THE GEOMETRIC PROPERTIES OF ELLIPSES, PARABOLAS AND HYPERBOLAS, J. W. Downs. This text shows how to create ellipses, parabolas, and hyperbolas. It also presents historical background on their ancient origins and describes the reflective properties and roles of curves in design applications. 1993 ed. 98 figures. xii+100pp. 6½ x 9¼. 0-486-42876-1

THE THIRTEEN BOOKS OF EUCLID'S ELEMENTS, translated with introduction and commentary by Sir Thomas L. Heath. Definitive edition. Textual and linguistic notes, mathematical analysis. 2,500 years of critical commentary. Unabridged. 1,414pp. 5⅜ x 8½. Three-vol. set.
Vol. I: 0-486-60088-2 Vol. II: 0-486-60089-0 Vol. III: 0-486-60090-4

SPACE AND GEOMETRY: IN THE LIGHT OF PHYSIOLOGICAL, PSYCHOLOGICAL AND PHYSICAL INQUIRY, Ernst Mach. Three essays by an eminent philosopher and scientist explore the nature, origin, and development of our concepts of space, with a distinctness and precision suitable for undergraduate students and other readers. 1906 ed. vi+148pp. 5⅜ x 8½. 0-486-43909-7

GEOMETRY OF COMPLEX NUMBERS, Hans Schwerdtfeger. Illuminating, widely praised book on analytic geometry of circles, the Moebius transformation, and two-dimensional non-Euclidean geometries. 200pp. 5⅝ x 8¼. 0-486-63830-8

DIFFERENTIAL GEOMETRY, Heinrich W. Guggenheimer. Local differential geometry as an application of advanced calculus and linear algebra. Curvature, transformation groups, surfaces, more. Exercises. 62 figures. 378pp. 5⅜ x 8½. 0-486-63433-7

History of Math

THE WORKS OF ARCHIMEDES, Archimedes (T. L. Heath, ed.). Topics include the famous problems of the ratio of the areas of a cylinder and an inscribed sphere; the measurement of a circle; the properties of conoids, spheroids, and spirals; and the quadrature of the parabola. Informative introduction. clxxxvi+326pp. 5⅜ x 8½. 0-486-42084-1

A SHORT ACCOUNT OF THE HISTORY OF MATHEMATICS, W. W. Rouse Ball. One of clearest, most authoritative surveys from the Egyptians and Phoenicians through 19th-century figures such as Grassman, Galois, Riemann. Fourth edition. 522pp. 5⅜ x 8½. 0-486-20630-0

THE HISTORY OF THE CALCULUS AND ITS CONCEPTUAL DEVELOPMENT, Carl B. Boyer. Origins in antiquity, medieval contributions, work of Newton, Leibniz, rigorous formulation. Treatment is verbal. 346pp. 5⅜ x 8½. 0-486-60509-4

THE HISTORICAL ROOTS OF ELEMENTARY MATHEMATICS, Lucas N. H. Bunt, Phillip S. Jones, and Jack D. Bedient. Fundamental underpinnings of modern arithmetic, algebra, geometry and number systems derived from ancient civilizations. 320pp. 5⅜ x 8½. 0-486-25563-8

A HISTORY OF MATHEMATICAL NOTATIONS, Florian Cajori. This classic study notes the first appearance of a mathematical symbol and its origin, the competition it encountered, its spread among writers in different countries, its rise to popularity, its eventual decline or ultimate survival. Original 1929 two-volume edition presented here in one volume. xxviii+820pp. 5⅜ x 8½. 0-486-67766-4

GAMES, GODS & GAMBLING: A HISTORY OF PROBABILITY AND STATISTICAL IDEAS, F. N. David. Episodes from the lives of Galileo, Fermat, Pascal, and others illustrate this fascinating account of the roots of mathematics. Features thought-provoking references to classics, archaeology, biography, poetry. 1962 edition. 304pp. 5⅜ x 8½. (Available in U.S. only.) 0-486-40023-9

OF MEN AND NUMBERS: THE STORY OF THE GREAT MATHEMATICIANS, Jane Muir. Fascinating accounts of the lives and accomplishments of history's greatest mathematical minds–Pythagoras, Descartes, Euler, Pascal, Cantor, many more. Anecdotal, illuminating. 30 diagrams. Bibliography. 256pp. 5⅜ x 8½. 0-486-28973-7

HISTORY OF MATHEMATICS, David E. Smith. Nontechnical survey from ancient Greece and Orient to late 19th century; evolution of arithmetic, geometry, trigonometry, calculating devices, algebra, the calculus. 362 illustrations. 1,355pp. 5⅜ x 8½. Two-vol. set. Vol. I: 0-486-20429-4 Vol. II: 0-486-20430-8

A CONCISE HISTORY OF MATHEMATICS, Dirk J. Struik. The best brief history of mathematics. Stresses origins and covers every major figure from ancient Near East to 19th century. 41 illustrations. 195pp. 5⅜ x 8½. 0-486-60255-9

Physics

OPTICAL RESONANCE AND TWO-LEVEL ATOMS, L. Allen and J. H. Eberly. Clear, comprehensive introduction to basic principles behind all quantum optical resonance phenomena. 53 illustrations. Preface. Index. 256pp. 5⅜ x 8½. 0-486-65533-4

QUANTUM THEORY, David Bohm. This advanced undergraduate-level text presents the quantum theory in terms of qualitative and imaginative concepts, followed by specific applications worked out in mathematical detail. Preface. Index. 655pp. 5⅜ x 8½. 0-486-65969-0

ATOMIC PHYSICS (8th EDITION), Max Born. Nobel laureate's lucid treatment of kinetic theory of gases, elementary particles, nuclear atom, wave-corpuscles, atomic structure and spectral lines, much more. Over 40 appendices, bibliography. 495pp. 5⅜ x 8½. 0-486-65984-4

A SOPHISTICATE'S PRIMER OF RELATIVITY, P. W. Bridgman. Geared toward readers already acquainted with special relativity, this book transcends the view of theory as a working tool to answer natural questions: What is a frame of reference? What is a "law of nature"? What is the role of the "observer"? Extensive treatment, written in terms accessible to those without a scientific background. 1983 ed. xlviii+172pp. 5⅜ x 8½. 0-486-42549-5

AN INTRODUCTION TO HAMILTONIAN OPTICS, H. A. Buchdahl. Detailed account of the Hamiltonian treatment of aberration theory in geometrical optics. Many classes of optical systems defined in terms of the symmetries they possess. Problems with detailed solutions. 1970 edition. xv + 360pp. 5⅜ x 8½. 0-486-67597-1

PRIMER OF QUANTUM MECHANICS, Marvin Chester. Introductory text examines the classical quantum bead on a track: its state and representations; operator eigenvalues; harmonic oscillator and bound bead in a symmetric force field; and bead in a spherical shell. Other topics include spin, matrices, and the structure of quantum mechanics; the simplest atom; indistinguishable particles; and stationary-state perturbation theory. 1992 ed. xiv+314pp. 6⅛ x 9¼. 0-486-42878-8

LECTURES ON QUANTUM MECHANICS, Paul A. M. Dirac. Four concise, brilliant lectures on mathematical methods in quantum mechanics from Nobel Prize-winning quantum pioneer build on idea of visualizing quantum theory through the use of classical mechanics. 96pp. 5⅜ x 8½. 0-486-41713-1

THIRTY YEARS THAT SHOOK PHYSICS: THE STORY OF QUANTUM THEORY, George Gamow. Lucid, accessible introduction to influential theory of energy and matter. Careful explanations of Dirac's anti-particles, Bohr's model of the atom, much more. 12 plates. Numerous drawings. 240pp. 5⅜ x 8½. 0-486-24895-X

ELECTRONIC STRUCTURE AND THE PROPERTIES OF SOLIDS: THE PHYSICS OF THE CHEMICAL BOND, Walter A. Harrison. Innovative text offers basic understanding of the electronic structure of covalent and ionic solids, simple metals, transition metals and their compounds. Problems. 1980 edition. 582pp. 6⅛ x 9¼. 0-486-66021-4

HYDRODYNAMIC AND HYDROMAGNETIC STABILITY, S. Chandrasekhar. Lucid examination of the Rayleigh-Benard problem; clear coverage of the theory of instabilities causing convection. 704pp. 5⅜ x 8¼. 0-486-64071-X

INVESTIGATIONS ON THE THEORY OF THE BROWNIAN MOVEMENT, Albert Einstein. Five papers (1905–8) investigating dynamics of Brownian motion and evolving elementary theory. Notes by R. Fürth. 122pp. 5⅜ x 8½. 0-486-60304-0

THE PHYSICS OF WAVES, William C. Elmore and Mark A. Heald. Unique overview of classical wave theory. Acoustics, optics, electromagnetic radiation, more. Ideal as classroom text or for self-study. Problems. 477pp. 5⅜ x 8½. 0-486-64926-1

GRAVITY, George Gamow. Distinguished physicist and teacher takes reader-friendly look at three scientists whose work unlocked many of the mysteries behind the laws of physics: Galileo, Newton, and Einstein. Most of the book focuses on Newton's ideas, with a concluding chapter on post-Einsteinian speculations concerning the relationship between gravity and other physical phenomena. 160pp. 5⅜ x 8½. 0-486-42563-0

PHYSICAL PRINCIPLES OF THE QUANTUM THEORY, Werner Heisenberg. Nobel Laureate discusses quantum theory, uncertainty, wave mechanics, work of Dirac, Schroedinger, Compton, Wilson, Einstein, etc. 184pp. 5⅜ x 8½. 0-486-60113-7

ATOMIC SPECTRA AND ATOMIC STRUCTURE, Gerhard Herzberg. One of best introductions; especially for specialist in other fields. Treatment is physical rather than mathematical. 80 illustrations. 257pp. 5⅜ x 8½. 0-486-60115-3

AN INTRODUCTION TO STATISTICAL THERMODYNAMICS, Terrell L. Hill. Excellent basic text offers wide-ranging coverage of quantum statistical mechanics, systems of interacting molecules, quantum statistics, more. 523pp. 5⅜ x 8½. 0-486-65242-4

THEORETICAL PHYSICS, Georg Joos, with Ira M. Freeman. Classic overview covers essential math, mechanics, electromagnetic theory, thermodynamics, quantum mechanics, nuclear physics, other topics. First paperback edition. xxiii + 885pp. 5⅜ x 8½. 0-486-65227-0

PROBLEMS AND SOLUTIONS IN QUANTUM CHEMISTRY AND PHYSICS, Charles S. Johnson, Jr. and Lee G. Pedersen. Unusually varied problems, detailed solutions in coverage of quantum mechanics, wave mechanics, angular momentum, molecular spectroscopy, more. 280 problems plus 139 supplementary exercises. 430pp. 6½ x 9¼. 0-486-65236-X

THEORETICAL SOLID STATE PHYSICS, Vol. 1: Perfect Lattices in Equilibrium; Vol. II: Non-Equilibrium and Disorder, William Jones and Norman H. March. Monumental reference work covers fundamental theory of equilibrium properties of perfect crystalline solids, non-equilibrium properties, defects and disordered systems. Appendices. Problems. Preface. Diagrams. Index. Bibliography. Total of 1,301pp. 5⅜ x 8½. Two volumes. Vol. I: 0-486-65015-4 Vol. II: 0-486-65016-2

WHAT IS RELATIVITY? L. D. Landau and G. B. Rumer. Written by a Nobel Prize physicist and his distinguished colleague, this compelling book explains the special theory of relativity to readers with no scientific background, using such familiar objects as trains, rulers, and clocks. 1960 ed. vi+72pp. 5⅜ x 8½. 0-486-42806-0